69 Structure and Bonding

Solid Materials

With Contributions by
J. Augustynski C. K. Jørgensen H. D. Lutz R. Reisfeld

With 32 Figures and 8 Tables

Springer-Verlag
Berlin Heidelberg GmbH

ISBN 978-3-662-15128-0

Library of Congress Cataloging-in-Publication Data. Solid materials. (Structure and bonding ; 69) Bibliography: p. Includes index. Contents: Aspects of photo-electrochemical and surface behaviour of titanium (IV) oxide / J. Augustynski – Excited state of chromium (III) in translucent glass-ceramics as prospective laser materials / R. Reisfeld, C. K. Jørgensen – Bonding and structure of water molecules in solid hydrates / H. D. Lutz. 1. Solid state chemistry. I. Augustynski, J. (Jan), 1941– . II. Series.   QD461.S 92 vol. 69 [QD478]   541.2′2s 88-3069
ISBN 978-3-662-15128-0   ISBN 978-3-540-48182-9 (eBook)
DOI 10.1007/978-3-540-48182-9
[541′.0421]

© Springer-Verlag Berlin Heidelberg 1988
Originally published by Springer-Verlag Berlin Heidelberg New York in 1988
Softcover reprint of the hardcover 1st edition 1988

Typesetting: Mitterweger Werksatz GmbH, 6831 Plankstadt, Germany

2152/3140-543210

Table of Contents

Aspects of Photo-Electrochemical and Surface Behaviour of Titanium(IV) Oxide

Jan Augustynski

Département de Chimie Minérale Analytique et Appliquée de l'Université, Quai Ernest Ansermet 30, CH-1211 Genève, Switzerland

Titanium(IV) oxide is continuing to attract wide interest as the photosensitizer for a large variety of photocatalytic reactions. Yet, the mechanisms of the photoreactions occurring at this semiconducting material are still poorly understood. In an attempt to clarify the existing ambiguities, the experimental evidence concerning the nature and role of the intermediates involved in the light-induced charge transfer reactions at TiO₂ is analyzed. Pathways for the photo-oxidation reactions, consistent with the available experimental data, are considered.

Structure and Bonding 69
© Springer-Verlag Berlin Heidelberg 1988

1 Introduction

The photoactivity of titanium dioxide is known since at least 65 years. Those first obser-vations, reported by Renz[1], concerned the fact that TiO_2 (and some other oxides), exposed to sunlight in the presence of glycerol or a carboxylic acid (for example, citric or tartaric acid), underwent darkening, causing simultaneously the photo-oxidation of the organic compound.

Among the early reports, regarding photocatalytic properties of titanium dioxide, one may also find a mention of its promoting effect upon the photo-oxidation of an inorganic compound ammonia[2].

A number of those early investigations were devoted to the "chalking" of paints containing white titania pigments[3]. This practically important phenomenon originates from the TiO_2-mediated photo-oxidation of the organic binder of a paint, producing carbon dioxide. Thus forming CO_2 gas film tends to separate the pigment grains from the binder, leading to the "chalking" of a TiO_2 powder off the paint.

The interest in the photochemistry of titanium dioxide has been greatly stimulated at the beginning of seventies by the works of Fujishima and Honda[4, 5]. The latter authors, using a single-crystal-rutile-TiO_2 photoanode, irradiated with near ultra-violet light, and a platinum cathode, have demonstrated the feasibility of the photoelectrolysis of water. A small external bias, much lower than that theoretically required in the case of conven-tional electrolysis, was sufficient to induce splitting of water into gaseous oxygen and hydrogen.

Although the device described by Fujishima and Honda operated with relatively high optical-to-chemical energy conversion efficiency, it was not well suited to the practically significant sunlight conversion. In fact, the band-gap energy of the n-type TiO_2 semicon-ductor (slightly higher for its anatase form, $E_g = 3.23$ eV, than for rutile, $E_g = 3.06$ eV) permits the direct absorption of only a small portion of the solar spectrum. This was the reason of numerous studies aimed at modifying the spectral photoresponse of titanium dioxide through doping[6-12]. The preference for using TiO_2 as a photoanode material is to be ascribed to its good stability (i.e., practical absence of the photocorrosion) as well as to the facility of preparing the polycrystalline films[6, 9]. The attempts to extend the photoresponse of titanium dioxide to the wavelengths greater than its fundamental absorption edge (i.e., ca. 400 nm) by incorporating foreign elements, such as chromium, nickel or manganese, have proven to be only partially successful. As a matter of fact, the presence of dopants of this kind in the TiO_2 lattice resulted usually in a rather drastic decrease of the quantum efficiency of the photocurrent[10]. Some cationic dopants, in particular those exhibiting very small ionic radii, as Be^{2+} and Al^{3+}, added to TiO_2 in relatively large amounts, permit a substantial improvement of the photocurrent-voltage characteristics of the polycrystalline electrodes, accompanied by the negative shift of the flat-band potential – without presenting any kind of drawback[6, 9]. Nevertheless, these dopants produce only marginal shift of the photoresponse towards visible light.

The sensitivity of titanium dioxide to visible light may be considerably improved by coating with dyes[13, 14]. The sensitized polycrystalline TiO_2 electrodes appear capable to sustain surprisingly high anodic photocurrents. For example, the photocurrent quantum efficiencies exceeding 40 per cent have recently been observed for anatase TiO_2 film electrodes coated with ruthenium tris(2,2'-bipyridil-4,4'-dicarboxylate) at the wavelengths ranging from 450 to 500 nm, i.e. close to the absorption maximum of the dye; these measurements were performed under slight anodic polarization of the electrode in a solution of pH 2.6 containing hydroquinone[14].

An alternative way to achieve the photodissociation of water consists in the use of aqueous suspensions of powdered or colloidal semiconductors, in general loaded with noble-metal and/or noble-metal-oxide catalysts[15–21], which act as short-circuited photoelectrolysis cells. Titanium dioxide was certainly (and is still being) the semiconductor most frequently employed in such systems.

Apart from water splitting, TiO_2 used in the form of particles or as a photoelectrode has been shown to catalyse also numerous other photoreactions. They include so different processes as, for example, photodegradation of chlorinated hydrocarbons[22–24] hydroxylation of aromatic hydrocarbons[25], photoreduction of dinitrogen[26–30] water-gas shift reaction[31, 32].

All photoreactions taking place at titanium dioxide and other n-type semiconductors are induced by the hole-electron pair formation subsequent to the light absorption. In principle, the holes and electrons, created by photons of an energy higher than the band-gap energy, become located in the valence, respectively, conduction band of the semiconductor. Their efficiency in driving photoreactions will depend upon relative rates of recombination and charge-transfer processes. In particular, the light to chemical conversion efficiency of the photochemical systems based on semiconductor suspensions will be determined by the competition between the photoanodic and photocathodic reactions, on the one hand, and the surface electron-hole recombination on the other. The above features are also expected to control the amount of the photocurrent at a semiconductor electrode in the region of potentials close to the flat-band potential. Under these conditions, the kinetics and, quite frequently, the very nature of the photoreactions occurring at titanium dioxide depend critically upon (i) the initial constitution of the TiO_2 surface, (ii) the modifications it undergoes during the band-gap irradiation and, in particular, the formation of more of less stable surface species constituting the reaction intermediates, (iii) the interactions between the titania surface and the ions and molecules present in the solution.

2 Interaction of TiO₂ with Water; the Surface Hydroxylation

Both the surface chemistry of titanium dioxide and the kinetics of photoreactions occurring at TiO_2 are markedly affected by its interaction with water.

There exists abundant experimental evidence, provided mainly by infra-red spectroscopy, that the TiO_2 surface, exposed to water vapour or to an aqueous solution, undergoes hydroxylation. The process of surface hydroxylation is thought to proceed through the chemisorption of molecular water on co-ordinatively unsaturated surface Ti^{4+} ions and its subsequent dissociation[33–35]. Due to the Lewis base character of the surface oxide ions, acting as donor surface states, the chemisorption of water at the Ti^{4+} site is likely to be followed by the proton transfer to the O^{2-} site resulting in the formation of two kinds of surface OH groups (Fig. 1). Such a picture is consistent with the calculations of surface adsorption energies of the water molecule on the rutile crystal, performed by Waldsax and Jaycock[36, 37]. These calculations included, in addition to the dispersion-repulsion

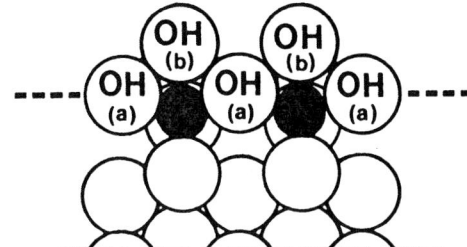

Fig. 1. Hydrated surface of anatase including basic (singly co-ordinated) and acidic
(doubly co-ordinated) OH groups[34, 137]

interactions, also the surface electrostatic field. For the (110) face of rutile, the most favourable adsorption site appears to be located over the five co-ordinated Ti^{4+} surface ion. Both the orientation of the water molecule – in the y-direction, with the dipole almost vertically downwards over the Ti^{4+} ion, and the computed adsorption energy, close to the first dissociation energy of H–OH, support the hypothesis of the dissociative adsorption:

$$_V[Ti^{4+}] + OH\text{–}H \rightarrow \,_{VI}[Ti^{3+\delta+}]OH^{\delta-} + H^+ \tag{1}$$

(the Roman numerals denote here the co-ordination numbers of the Ti atom). In the absence of other compounds having strong basic character, the proton, proceeding down the steepest field slope, is likely to react with a vicinal O^{2-} ion, forming a second kind of OH group:

$$_V[Ti^{4+}] + OH\text{–}H + \,_{II}[O^{2-}] \rightarrow \,_{VI}[Ti^{3+\delta+}]OH_b^{\delta-} + OH_a^- \tag{2}$$

The presence of δ in Eqs. (1) and (2) implies that the resultant bonding is not totally ionic[36].

The doubly-bonded (bridged) OH groups, formed according to Eq. (2), become thus Brönsted acid centers, while the singly-bonded (terminal) hydroxyls are expected to exhibit a predominantly basic character[34, 38]. A large amount of reliable data regarding the interaction of water with titanium dioxide, are provided by infra-red spectroscopic studies associated with temperature-programmed desorption measurements.

2.1 Thermo-Analytical and Infra-Red Spectroscopic Studies

Temperature-programmed desorption (TPD) measurements indicate, for both rutile and anatase TiO_2, the presence of at least two forms of chemisorbed water. For example, the TPD profiles associated with the dehydration of a powder rutile sample give rise to two peaks located at 643 K and, respectively, 523 K[39]. Unlike the higher temperature peak, the lower temperature desorption maximum was observed to be affected by the amount of adsorbed water and to be shifted from 593 K down to 523 K with increasing the coverage. On the basis of a comparison with infra-red spectra of a hydroxylated rutile surface[40], Munuera and Stone[39] assigned the TPD peak at 643 K to the dissociatively adsorbed water (cf. Eqs. (1) and (2)) and the lower temperature TPD maximum to the chemisorbed molecular H_2O. An analogous investigation, involving thermogravimetric analysis (TGA), temperature-programmed desorption and infra-red (IR) spectroscopy, has more recently been performed by Munuera et al.[41] with an anatase P25 sample (made by Degussa in Frankfurt) extensively used in the study of different photocatalytic reactions (the P25 TiO_2 contains also some amount of rutile, possibly present on the very surface of the particles). The three peaks, observed in the derivatographic thermog-ravimetric curves at ~ 373, 488 and 598 K, were attributed, respectively, to loosely bound (physically adsorbed) water, to the chemisorbed molecular water and, for the DTGA peak at 598 K, to the hydroxyl groups. This assignment is consistent with the results of IR spectroscopic analyses of the anatase P25 samples outgassed at increasing temperatures ranging from 373 to 673 K[41]. Thus, the absorption band at 1610 cm^{-1}, due to molecular water, still present in the sample outgassed at 423 K, was no more visible after pretreatment above 473 K. The IR spectra of the anatase samples outgassed at higher temperatures showed absorption exclusively in the OH stretching region with a strong band at 3730 cm^{-1} and a series of weaker bands at 3680, 3620 and 3580 cm^{-1}. These stretching frequencies are in reasonable agreement with those reported earlier by Yates[42] for anatase and by Primet et al.[43, 44] in their comparative IR study of hydroxy-lated anatase and rutile powders. The latter authors assigned the high-frequency absorp-tion bands, observed at 3715 cm^{-1} for anatase and at 3685 cm^{-1} for rutile, to the isolated OH groups, and the remaining lower-frequency bands to the OH groups bonded by hydrogen bridges. With progressively increasing the pretreatment (outgassing) tempera-ture of the sample, the high-frequency bands for anatase and rutile were the last to desappear from the IR spectra. This is consistent with the behaviour expected for the isolated OH groups situated less favourably for removal from the surface than the hydrogen-bonded hydroxyls[45]. Primet et al.[44] have also made interesting observations regarding largely irreversible nature of the dehydroxylation process. They have found both anatase and rutile samples to undergo the initial dehydroxylation by outgassing at 623–673 K. Repeating the dehydroxylation-rehydroxylation treatments resulted in a pro-found modification of the surface and a concomitant decrease of the dehydroxylation temperature. The IR spectra, recorded for the TiO_2 sample subjected to ten cycles of this kind, exhibited only broad bands characteristic of the hydrogen-bonded OH groups which were removed by evacuation at already 423 K. The above changes in the energe-tics of the water adsorption and, in particular, the suppression of the initially observed isolated OH groups could not be simply explained by the decrease in the surface area of the TiO_2 during the dehydroxylation-rehydroxylation cycles, which did not exceed 20 per cent after ten cycles. Primet et al.[44] suggested the formation of an increasing number of incompletely co-ordinated titanium atoms, occurring during the cyclic treatments, to be the possible reason for the limited rehydroxylation. The incompletely co-ordinated sur-face Ti atoms are expected to act as electron-acceptor sites and, consequently, to favour the adsorption of electron-donor molecules[44]. On the other hand, such restructured TiO_2 surface should be much less propitious to the adsorption of the species which, like oxygen, tend to act themselves as electron acceptors.

The infra-red studies provide also some insight into the effect of impurities on the hydroxylation of the TiO_2 surface. The importance of these effects is easily appreciated if one takes into account that both the main routes (the oxidation or hydrolysis of titanium tetrachloride and the hydrolysis of titanium sulphate) used for the preparation of TiO_2 samples are likely to leave anionic impurities at the surface. The residual chloride present on the surface of rutile has been shown by Jackson and Parfitt[35] to affect significantly the

OH stretching region of the IR spectra. Also the dehydroxylation of the TiO_2 surface containing chloride impurity occurs apparently at a lower temperature than for a chloride-free surface[46]. Chloride ions, migrating to the surface during heating, are thought to assist in the removal of the OH species[46].

It seems quite likely that not only the surface properties but also the (photo)catalytic behaviour of TiO_2 may be affected by the presence of chloride ions in the lattice of the oxide. The latter is practically unavoidable in the case of TiO_2 deposits, originated from the hydrolysis (oxidation) of titanium tetrachloride – frequently used in photoelectrochemical studies. In fact, the methods enabling removal of the surface chloride from the TiO_2 powders (Soxhlet extraction with water and heating in air at high temperature)[35] are not applicable to other kinds of TiO_2 samples.

2.2 Electron Spectroscopic Studies

An attempt of identification of the two kinds of OH groups on the surface of a single-crystal (the (001) face) of rutile has been undertaken by Sham and Lazarus[47] using X-ray photoelectron spectroscopy (XPS). Modulating the effective sampling depth by positioning the analyser at different angles with respect to the sample surface, the latter authors were able to resolve the O 1s photoelectron spectrum of TiO_2 into the contributions arising from the bulk (under-surface) lattice O^{2-} ions and from the very surface oxygens. In the case of an hydrated rutile surface, the latter contribution, represented by the higher binding energy shoulder of the O 1s signal, was resolved into two peaks, at ca. 532.2 and 533.6 eV, assigned to acidic, respectively, basic OH groups. This assignment, however, does not take into account the presence at the rutile surface of chemisorbed molecular water.

The comparison of the above XPS results[47] with those of an extensive surface study of the (100) plane of rutile, performed by Lo et al.[48] using ultraviolet photoemission spectroscopy (UPS), electron-energy loss spectroscopy (ELS), Auger electron spectroscopy (AES), low-energy electron diffraction (LEED) and thermal desorption, reveals an apparent disagreement regarding the surface conditions of TiO_2 under which the dissociative adsorption of water may take place. From the observed differences in the UPS and ELS spectra, recorded for the stoichiometric and, respectively, reduced TiO_2 rutile surface, the latter authors have concluded that the presence of Ti^{3+} surface species is a necessary prerequisite for the dissociative adsorption of water. The photoemission difference, $\Delta N(E)$, curves for chemisorption of water on the stoichiometric (100) rutile surface, obtained by Lo et al.[48], were very similar to the ultraviolet photoemission spectrum of gas phase water, in contrast with the markedly changed $\Delta N(E)$ curves for the Ti^{3+}-rich surfaces – indicating that the adsorbate was no more molecular water but, as suggested, the hydroxyl groups. In the latter case the chemisorption of water caused distinct decrease of intensity of the -0.6 eV emission in UPS, associated with the Ti^{3+} ions, indicating a strong interaction between the latter species and the adsorbate. Importantly, the band-gap (>3 eV) illumination of the TiO_2 surface was shown to lead to a regeneration of the initial amount of the Ti^{3+} species.

On the basis of these observations, Lo et al.[48] have postulated that the Ti^{3+} ions, present on the surface of a TiO_2 photoanode, play an essential role in the process of photo-oxidation of water and that, on the other hand, the stoichiometric TiO_2 should be inactive in photodecomposing water because of the lack of suitable sites for the dissociative H_2O adsorption.

The results of the work of Lo et al.[48] seem, at first sight, difficult to conciliate with the extensive experimental evidence, arising from different IR spectroscopic studies[35, 42, 44, 49–52], indicating that the OH groups do form on the surface of unreduced TiO_2 powders exposed to water vapour. It is necessary to recall in this connection that, according to Jones and Hockey[53], the geometry of both the (101) and (100) planes of rutile does not permit the dissociative chemisorption of water, in contrast to the (110) plane on which the dissociation of adsorbed water appears to be allowed. As the external surface of the fine crystallites constituting the rutile powders is frequently assumed to be composed principally of three planes: (110), (101) and (100) accounting, respectively, for 60, 20 and 20 per cent of the total surface area[53], a significant extent of the dissociative H_2O adsorption is to be expected for the powder samples. In the light of the considerations of Jones and Hockey[53] also the results of Lo et al.[48], obtained for the (100) face of a single crystal of rutile, become understandable, the latter plane being assumed inactive with regard to the dissociation of adsorbed water. In this connection, an extension of the surface spectroscopic studies, principally to the (110) plane of rutile, would be of evident interest as a mean of definitive testing the Jones and Hockey's hypothesis regarding different adsorption behaviour of various planes of the rutile crystal.

Assuming, following the above mentioned suggestion of Lo et al.[48], that the activity of the TiO_2 surface for photodecomposing water should be considerably affected by its ability to adsorb H_2O dissociatively, one may expect significant differences between the photoactivities of the (110), respectively, (100) plane of rutile. In particular, the photoactivity of the former should be much less influenced by the extent of the surface reduction than that of the latter, requiring large surface concentration of the Ti^{3+} species to act as sites for the initial dissociative adsorption process.

2.3 Nuclear Magnetic Resonance Spectroscopy of the TiO₂ Surface

Enriquez et al.[54] have investigated the constitutive water (i.e., the OH groups and H_2O molecules) present on the surface of anatase, using ^1H nuclear magnetic resonance (NMR). The samples were the portions of a high-surface-area synthetic anatase powder, outgassed at a series of temperatures ranging from 295 to 668 K. The spatial distribution of protons between the hydroxyl groups and the water molecules was obtained from the calculated shape functions, simulating the experimental spectra. In the model assumed by Enriquez et al.[54], the magnetic interactions between the protons of the OH groups and of the H_2O molecules were represented by a three-spin configuration, consisting of two protons of a water molecule, forming the base of an isosceles triangle, and of a hydrogen atom of a hydroxyl group, placed at the third corner. In addition, they considered a two-spin configuration to describe the excess of OH groups or H_2O molecules on the surface.

The analysis of the NMR spectra of the samples with different extents of hydration, according to the above model, enabled to calculate the coverages of the hydroxyl groups and of the molecular water. For the less dehydrated sample, approximately half of the total amount of 9 equivalent water molecules per square nanometer of the surface was found to be present in the form of the OH groups. The resulting concentration of ca.

9 OH groups/nm^2 was considered to equate with a complete monolayer[1], thus excluding the possibility of the direct bonding of molecular water to the Ti^{4+} sites on the surface. Accordingly, the remaining water molecules, ca. 4.5 H_2O/nm^2, were postulated to form an overlayer, attached by hydrogen bonds to the doubly co-ordinated (acidic) OH groups.

On the basis of the results obtained for the samples with varying extents of hydration, the dehydration appears to proceed through the simultaneous removal of molecular water and of a restricted number of hydroxyl groups. In addition, the final loss of water was associated with the broadening of the NMR spectra, observed for the relatively dehydrated samples, explained by the appearance of the paramagnetic Ti^{3+} sites.

2.4 General Remarks

From the above discussed results of various experimental studies emerges a comprehensive picture of the hydrated TiO_2 surface covered, in addition to the two kinds of hydroxyl groups, by the chemisorbed molecular water and the physically adsorbed water. There are, however, certain differences with regard to the distribution of these species and the way by which they are bound to the sites on the surface. There is rather ample agreement that not more than half of the surface of anatase and rutile powders undergoes hydroxylation through the dissociative adsorption of water. This view is consistent with the adsorption isotherms of various compounds, known to react with the surface hydroxyl groups[34, 38], as well as with the results of TPD experiments[39]. Accordingly, the concentration of the surface hydroxyls, determined for a large variety of polycrystalline anatase and rutile samples, ranges from ca. 4.5 to 5 OH groups/nm^2. In comparison, the value of ca. 9.3 OH/nm^2, calculated by Enriquez et al.[54] from their NMR data, appears surprisingly high. However, the unusually large surface area (320 m^2/g) of the anatase employed in that study suggests a possible presence of micropores in its surface. In such a case, the potential-determining ions might be accomodated not only on the surface plane of the TiO_2 particles but also inside the micropores[2] leading to an abnormally high OH coverage.

The largely prevailing view is that, besides the hydroxyl groups, also the molecular water is bound directly to the co-ordinatively unsaturated surface Ti^{4+} ions. Combined TPD and IR data indicate that there are 2 to 3 molecules per nm^2 of such strongly adsorbed water on the surface of TiO_2[39]. The removal of this chemisorbed water from the surface of rutile requires, in general, the heating above 470 K[51].

The mode (or modes) of bonding of the second kind (weakly adsorbed) of molecular water, removed by outgassing at room temperature, appears less clear. Hydrogen bonding to the surface OH groups[51] and adsorption at the isolated surface O^{2-} ions[39] have been alternatively suggested.

---

1 The above estimate was based upon comparison with the theoretical size of an OH group located on the surface of a solid, taken as 0.127 nm^2 [55]. This implies, however, that, under the saturation coverage, a significant amount of surface sites remain unoccupied. For example, in the case of the (001) plane of anatase, the portion of unoccupied sites would be of 33 per cent[54].
2 The concept of permeability of the oxide surface to the ions, known as the porous double-layer model, has originally been advanced to explain extremely large surface charges observed with some SiO_2 samples[56, 57].

According to the model proposed by Jones and Hockey[53], the strong adsorption, i.e., the co-ordinative bonding of water would be restricted to the (100) and (101) planes of rutile, while the weak adsorption, through the hydrogen bonding to the OH groups, should be expected to occur on the (110) plane. Assuming that (i) the external surface of rutile powders consists of the (110), (100) and (101) planes in a ratio 3 : 1 : 1, and that (ii) only the (110) plane undergoes hydroxylation, Jones and Hockey[53] have obtained, from the corresponding lattice parameters, the theoretical surface density of 6.12 OH groups and 3.06 co-ordinatively bound H_2O molecules per nm^2. These values are, in fact, only slightly larger than those determined experimentally.

The above considerations are applicable to the titanium dioxide samples prepared and pretreated following the standard procedures. The presence of significant amounts of impurities or, for example, the high-temperature pretreatment of the TiO_2 samples lead usually to a more or less important decrease of the extent of hydroxylation.

There are several indications that the reduction of the TiO_2 surface weakens its interaction with water[58–60]. Such observations are important with respect to the photo-electrochemical behaviour of titanium dioxide since the preparation of the TiO_2 photo-electrodes or photocatalysts involves, in general, a reduction pretreatment intended to increase the donor concentration.

In particular, Iwaki and Miura[60] have ascribed the observed decrease of the heat of immersion in water of an anatase sample, reduced by hydrogen at 873 K, to the partial removal of the surface O^{2-} ions. They have also noted that the normal interaction with water (typical of unreduced TiO_2) was not restored after the reoxidation of the sample at 873 K. The question arises, however, whether such behaviour is associated with that particular mode of TiO_2 reduction or is also characteristic of the samples reduced by other means.

Of direct interest with regard to the photoelectrochemistry of TiO_2 is also the report of Misra[61] indicating that the irradiation of an anatase sample with the UV light causes dramatic decrease of the temperature required for the dehydroxylation. Thus, the irradi-ated TiO_2 (anatase) samples were observed to undergo dehydroxylation by outgassing at temperatures not exceeding 423 K, instead of the usual temperature of 823 K.

Some observations, relevant to the effect of the UV illumination upon energetics of water adsorption on the (100) plane of rutile TiO_2, have been made by Lo et al.[48]. The band-gap (hv > 3 eV) irradiation of the prereduced, Ti^{3+}-rich rutile surface, containing adsorbed water, resulted in a distinct increase of the work function, indicating a possible change of the nature of adsorbed species.

The above data, even if fragmentary, suggest that a prolonged, intense UV illumina-tion may eventually affect the extent of hydroxylation of titanium dioxide and of other semiconducting oxides. This kind of effects could probably be checked directly by the potentiometric titrations of the illuminated and unilluminated oxide suspensions.

3 Interface Between TiO$_2$ and Aqueous Solutions

3.1 Surface Ionization Reactions and Point of Zero Charge (PZC) of TiO$_2$

The amphoteric character of the hydroxylated TiO$_2$ surface has been demonstrated by Boehm and co-workers[33, 34, 38, 62] in a series of potentiometric titration experiments performed both with anatase and rutile samples. Approximately half of the total amount of hydroxyls present on the TiO$_2$ surface, which underwent neutralization with a diluted (0.01 M) sodium hydroxide solution, has been described as relatively strongly acidic, with a pK value of 2.9, and the remainder as weakly acidic with a pK of 12.7[34]. However, the use of a Langmuir-type adsorption equation as a basis for estimating these acid dissociation constants has raised questions about the real significance of the above values[63].

The amphoteric ionization reactions of the surface OH groups[64]:

$$Ti_s - OH \rightleftharpoons Ti_s - O^- + H_{aq}^+ \tag{3}$$

$$Ti_s - OH + H_{aq}^+ \rightleftharpoons Ti_s - OH_2^+ \tag{4}$$

determine the surface charge of the oxide-solution interface. In practice, the surface charge density, σ_s, is usually calculated from the uptake of the potential-determining ions, H$^+$ and OH$^-$, measured as a function of the pH of the oxide suspension:

$$\sigma_s = F(\Gamma_{H^+} - \Gamma_{OH^-}) \tag{5}$$

where Γ_{H^+} and Γ_{OH^-} are the surface excesses (the adsorption densities) of the corresponding ions, expressed in mol · cm^{-2}, and F is the Faraday constant.

The pH value at which the oxide surface carries no fixed charge, i.e. $\sigma_s = 0$, is defined as the point of zero charge (PZC)[65]. A closely related parameter, the isoelectric point (IEP), obtained from electrophoretic mobility and streaming potential data, refers to the pH value at which the electrokinetic potential equals to zero[65]. The PZC and IEP should coincide when there is no specific adsorption in the inner region of the electric double layer at the oxide-solution interface. In the presence of the specific adsorption, the PZC and IEP values move in opposite directions as the concentration of supporting electrolyte is increased[66–68].

Apart from the specific adsorption, the value of the PZC may also be affected by the titration rate[69]. In fact, as shown by the potentiometric titration experiments on various oxides[70–74], including TiO$_2$[73], the uptake of the potential-determining ions is a two-step process. The first adsorption step is fast, being completed in few minutes, while the second step may require several weeks to reach equilibrium[75]. For this reason, rapid titration procedures, allowing to avoid the slow adsorption step, are generally considered to yield more significant PZC values[75]. An example of such a procedure is furnished by the method developed by Ahmed[76], according to which the surface charge is calculated from the initial changes of pH, observed upon immersion of dry samples of oxide in solutions having different pH values. This method enables also to minimize the eventual effects resulting from the dissolution of the oxide.

Nevertheless, the observation of the slow adsorption processes involving the potential-determining ions, which are expected to reflect slow changes in the PZC[69, 75], may be of interest in relation to long-term photochemical or photoelectrochemical experiments with TiO_2 dispersions or TiO_2 electrodes.

Recent careful measurements of surface charge density of titanium dioxide (rutile) suspensions as a function of pH, performed by Yates and Healy[75], employing aqueous KNO_3 solutions of various concentrations, place the PZC at pH = 5.8 ± 0.1. The sample used in this study was a thoroughly purified synthetic rutile prepared by hydrolysis of titanium tetrachloride. The fact that Wiese and Healy[77] determined, for a similar rutile sample in KNO_3 solution, the IEP value of 5.8 (i.e. equal to the PZC) would indicate that potassium nitrate behaves as a classical "indifferent electrolyte" and that the above PZC and IEP of TiO_2 may be considered as intrinsic values. A slightly higher value of the PZC (pH = 6) has earlier been reported by Bérubé and de Bruyn[73] for a rutile sample (obtained also *via* the chloride route) immersed in a NaCl supporting electrolyte.

The available data regarding the point of zero charge of anatase are not markedly different from those for the rutile and range from pH = 6 to pH = 6.4[73, 78, 79].

There is abundant evidence, especially in earlier literature[64], that the values of the PZC and IEP are strongly affected by a high-temperature pretreatment of the oxide. In fact, calcining of the TiO_2 sample above 1073 K (in general, at about 1273 K) results in dropping of the PZC and IEP to pH 4.7–4.8[80, 81]. Such a decrease of the PZC/IEP values, exceeding one unity of pH, reflects apparently irreversible changes in the surface chemistry of titanium dioxide. It is significant, in this connection, that the TiO_2 samples, after being calcined above 1073 K, undergo only partial rehydroxylation when exposed to water[40, 82].

3.2 Adsorption of Inorganic Ions on the Surface of TiO_2

Surface charge density (σ_s) versus pH isotherms, derived from the results of potentiometric titrations, are also an important source of information about the ions adsorption on the oxide surfaces. Comparison of the σ_s against pH curves obtained in solutions of different electrolytes allowed to establish sequences of preferential adsorption of cations on titanium dioxide as being: $Mg^{2+} > Li^+ > K^+ > Cs^+$, and for anions: $Cl^- \simeq ClO_4^- \simeq NO_3^- \simeq I^-$ [75, 83]. The occurrence of the specific adsorption³ of a cation is reflected by the changes in the negative branch of the σ_s-pH curve, reaching much higher absolute σ_s values than those observed in the solution of an "indifferent electrolyte". As already mentioned, aqueous KNO_3 behaves with respect to the TiO_2 as a practically "indifferent electrolyte", giving rise to a relatively symmetrical σ_s-pH curve[75]. Much larger surface charge densities of TiO_2, determined in $LiNO_3$ and, especially, $Mg(NO_3)_2$ solutions, at the pH values higher than the PZC, are apparently indicative of the specific adsorption of Li^+ and Mg^{2+} ions[75]. In addition, in concentrated (e.g., $1 \ mol \cdot dm^{-3}$) $LiNO_3$ solutions the PZC of titanium dioxide shifts towards lower pH values[75].

---

3 The term "specific adsorption" is here employed as synonymous with "super-equivalent adsorption"[86]. The occurrence of the super-equivalent adsorption of an ion is usually revealed by the pronounced asymmetry of the σ_s against pH curve and a shift of the PZC.

The requirement of ion adsorption to balance the charges on the oxide surface has been formulated in the site-dissociation/site-binding model developed by Yates, Levine and Healy[84, 85] and by Davis, James and Leckie[87, 88]. According to this model, charging of the oxide surface brings about binding of the counter ions (from the supporting electrolyte) to the ionized OH groups. The whole process is thus described by the combination of the amphoteric ionization reactions (Eqs. (3) and (4)) with two small ion-binding equilibria[84]:

$$Ti_s - OH_2^+ + A_{aq}^- \rightleftharpoons Ti_s - OH_2^+ A_{aq}^- \tag{6}$$

$$Ti_s - O^- + K_{aq}^+ \rightleftharpoons Ti_s - O^- K_{aq}^+ \tag{7}$$

Here, K_{aq}^+ and A_{aq}^- denote, respectively, the cation and anion of the supporting electrolyte, and the sign aq is intended to imply that the ions undergoing adsorption in that way preserve virtually their hydration shell.

In practice, the surface charge of the TiO_2 dispersions formed in very dilute salt solutions is mainly determined by the amphoteric ionization reactions, Eqs. (3) and (4), while, with increasing concentration of the supporting electrolyte, the ion-binding reactions, Eqs. (6) and (7), become predominant. Unlike the ionized surface hydroxyl groups, $Ti_s - OH_2^+$ and $Ti_s - O^-$, the dipolar $Ti_s - OH_2^+ A^-$ and $Ti_s - O^- K^+$ groups are expected not to contribute to the charge of the diffuse part of the double layer.

The site-dissociation/site-binding model[84, 87] enables to account for the large surface charge densities, observed for most oxides, without assuming any special structure of the surface layer. Other models, such as the porous double layer model[56, 57, 89] and the gel layer model[90–93], seem applicable mainly to some particular kinds of oxides associating extremely high surface charge densities with high surface porosities.

An important observation made by Bérubé and de Bruyn[83], providing some insight into the nature of the interactions between the surface of titanium dioxide and the solution, was that the sequence of preferential adsorption of cations on TiO_2 is the reverse of the order found, e.g., for mercury[94], silver iodide[95] and also for silicon dioxide[56]. The ions preferentially adsorbed on titanium dioxide are those with strong hydration tendencies and marked structure-promoting aptitudes, such as small Mg^{2+} and Li^1 cations. Although the anions examined by Bérubé and de Bruyn[73, 83] showed, in general, little specific affinity for the rutile surface, the adsorbability of the I^- ion appeared still lower than those of Cl^-, ClO_4^- and NO_3^-.

Similar observations, regarding the adsorption sequences of cations and anions, have also been made for other oxides (α-Fe_2O_3, ZnO) and have led to a description of such oxide surfaces as structure promoting[83, 96]. This representation constitutes an extension of the Gurney's[97] interpretation of ion-ion interactions in solution to the ion-surface interactions.

The term which, in the light of the site-dissociation/site-binding model, appears as rather ambiguous is that of the specifically adsorbable ion. In fact, the specific (super-equivalent) adsorption of an anion or that, less frequent, of a cation at the metal-solution interface involves the displacement of the hydration shell around the ion[4], allowing its close interaction with the metal surface[98]. More exactly, the chemisorbed ion is regarded

---

4 The specifically adsorbed ion is, however, regarded as still partly hydrated.

as a "ligand" to the surface, forming a co-ordinate bond with the metal[99, 100]. A number of specifically adsorbed anions are actually considered to undergo partial electron charge transfer towards the metal, as, for example, the I^- ion adsorbed at platinum.

In contrast with the situation at the metal surfaces, it is not clear whether the specific (super-equivalent) ion adsorption at the oxide-solution interface involves a qualitative change in the nature of the bonding to the surface, with respect to the equivalent counter-ion adsorption. The fact that the tendency of cations for the preferential adsorption on TiO_2 increases markedly with increasing the strength of hydration rather precludes the eventuality of the close bonding of the (significantly dehydrated) cations to the discrete charges on the oxide surface.

The factor which may explain the aptitude of the high charge-density ions, such as Mg^{2+} and Li^+, for the super-equivalent adsorption at TiO_2 is their ability to interact sufficiently strongly with the co-ordinating water molecule to weaken the O–H bonds. In other words, a cation of small radius, such as Li^+, is expected to polarize a water molecule

$$Li^+ \longleftarrow O \overset{\displaystyle /H^{\delta+}}{\underset{\displaystyle \backslash H^{\delta+}}{}}$$

enhancing the acidic character of its hydrogen atom[101]. The resulting "localized hydrolysis" may facilitate the attachment of cations of this kind also to the non-ionized (basic) OH groups on the TiO_2 surface, through the formation of the hydrogen bond:

$$Ti_s\text{–O–H} + \overset{\displaystyle H\backslash}{\underset{\displaystyle H\nearrow}{O}} \longrightarrow Li^+ \longrightarrow Ti_s\text{–O–H}\text{--}\underset{\displaystyle |}{\overset{\displaystyle O^-}{\underset{\displaystyle H}{}}} \longrightarrow Li^+ + H^+_{aq} \qquad (8)$$

The latter mechanism, operating in addition to the equivalent adsorption of cations on the ionized $Ti_s - O^-$ sites (cf. Eq. (7)), would account for the abnormally large negative surface charge densities, observed in solutions containing Mg^{2+} or Li^+ ions, at pH's higher than the PZC.

The specific adsorption of anions at titanium dioxide (discussed in the following part of this chapter) seems to occur via the replacement of the basic OH groups[34]. It appears difficult, at the first sight, to find a common point among the anions known to give rise to the super-equivalent adsorption on the surface of TiO_2.

The interactions between the anions and the TiO_2 surface have also been investigated by other, more direct techniques than the potentiometric titrations. Thus, Sanchez and Augustynski[102, 103] have employed X-ray photoelectron spectroscopy to examine the adsorption of a series of anions onto titanium dioxide films of two kinds: the relatively thick (10–15 μm) films, obtained by thermal decomposition of an alcoholic $TiCl_4$ solution, and the natural, thin (15–20 Å) films covering the surface of titanium metal. The TiO_2 films of the first kind, supported onto Ti metal, have been extensively used as electrode materials in electrochemical[104] and photoelectrochemical studies[9].

The adsorption experiments consisted in a prolonged exposure of the samples to the aqueous solutions of different sodium salts of pH $\approx$ 6, except for the phosphate buffer solutions of pH $\approx$ 7. This enabled to avoid significant pH changes during washing of the samples in doubly distilled water. Before being transferred to the spectrometer, the TiO_2 samples were dried in a desiccator. Such pretreatment is expected to remove from the TiO_2 surface the physically adsorbed water and, possibly, a part of the chemisorbed water, preserving, however, the surface OH groups. The latter point was confirmed by the presence, in the O 1 s photoelectron spectra, of an additional, higher-binding-energy signal attributable to the OH species and/or to the chemisorbed water[102].

Under so chosen experimental conditions, only the anions undergoing irreversible (non-equilibrium) adsorption on titanium dioxide, in the range of pH close to or slightly higher than the PZC, should be detectable by the XPS analyses. The results of these measurements, summarized in Table 1, indicate that three anions, $H_2PO_4^-$ and/or HPO_4^{2-}, F^- and NO_3^- interact particularly strongly with TiO_2.

As the mean escape depth of the photoelectrons originating from the outermost layer of TiO_2 (with kinetic energies ranging from 700 to 800 eV) does not exceed 15–20 Å, the observed relative amounts of fluorine, nitrogen and phosphorus species correspond to high coverage of the TiO_2 surface, approaching half a mono-layer.

It is to be mentioned, in this connection, that the chemisorbed $H_2PO_4^-$ [105, 106] and F^- [107] ions have been reported to affect the photo-oxidation reactions at TiO_2 and to be able to replace the basic (singly-bonded) hydroxyl groups on the surface of TiO_2[108–111]. The observations, regarding a strong decrease of the extent of CO_2 adsorption on TiO_2 pretreated with phosphoric acid, have been interpreted in this sense[34].

A mechanism involving direct bonding of anions with the co-ordinatively unsaturated surface Ti^{4+} ions, in replacement of the basic OH groups, offers a plausible explanation of the irreversible character of adsorption of the $H_2PO_4^-$ and F^- ions. There exists some evidence that other anions, such as, for example, $PtCl_6^{2-}$ and/or $PtCl_4^{2-}$, CH_3COO^- [112] or $HCOO^-$ may also be adsorbed on TiO_2 according to the same mechanism. Recently, also arsenate ions have been reported[113] to undergo strong adsorption on hydrous TiO_2 in the way similar to the phosphate ions.

This kind of irreversible adsorption has to be clearly distinguished from the above discussed site-binding mechanism (cf. Eq. (6)) accounting for the equilibrium adsorption of such anions as Cl^-, I^- or SO_4^{2-}.

The experiments performed by Sanchez and Augustynski[102] have also revealed that the adsorption of oxy-anions on the TiO_2 films may be accompanied by their reduction.

Table 1. Extents of adsorption[a] of the anions identified on two kinds of TiO_2 films pre-exposed to various solutions; for comparison, some results obtained with a commercial anatase powder are represented in the last column[102, 103, 114]

Solution	Thin (15–20 Å) air-formed TiO_2 film	Thick (10–15 μm) TiO_2 deposit (anatase)	TiO_2 (anatase) powder (Merck)
0.1 M NaCl	not detected	not detected	
1 M NaCl	not detected		
0.1 M NaClO$_4$	~4 (as Cl$^-$)	2 (as Cl$^-$)	
1 M NaClO$_4$	~3 (as Cl$^-$)		
1 M NaNO$_3$	15 (as NH$_4^+$)	9 (as NH$_4^+$)	not detected
0.1 M NaH$_2$PO$_4$/Na$_2$HPO$_4$	12	5	8
1 M NaH$_2$PO$_4$/Na$_2$HPO$_4$	8		
0.1 M NaF	9	8	7
0.1 M Na$_2$SO$_4$	not detected	not detected	

[a] Expressed in atom percent relative to Ti(IV) of TiO_2, taken as 100 at.%

Thus, the species, present on the surface of the samples pre-exposed to sodium nitrate solutions, have been unambiguously identified, on the basis of the N 1s photoelectron spectra, as ammonia or, possibly, NH_4^+ ions. This result appears less unexpected in the case of thin, natural TiO_2 films, covering the Ti metal[5], than in that of the 10–15 μm thick films. There are, presumably, the Ti^{3+} ions, present in the surface region of the latter TiO_2 films, which are responsible for the observed reduction of the nitrate ions. This view is consistent with the fact that no nitrogen species (i.e., neither the NO_3^- ions nor their reduced form) were detected in the similar XPS experiments performed with an unreduced commercial anatase powder exposed to a $NaNO_3$ solution[114]. Thus, the reduction of the NO_3^- anions appears as a prerequisite for their irreversible adsorption on TiO_2. On the other hand, the extents of adsorption of the $H_2PO_4^-$ (HPO_4^{2-}) and F^- ions, on both kinds of TiO_2 films and on the unreduced TiO_2 powder, were practically identical.

The perchlorate ions have been found to interact with the TiO_2 films (but not with the unreduced anatase powder), forming relatively small amounts of chloride species (cf. Table 1). It seems probable that the latter species, arising from the "reductive adsorption" of the ClO_4^- ions, were not simply adsorbed but incorporated into the surface region of the TiO_2 films, as no irreversible chloride adsorption could be detected from NaCl solutions.

Kazarinov et al.[117] have investigated the adsorption of ClO_4^-, Cl^-, HSO_4^- and $H_2PO_4^-$ ions on titanium dioxide by means of the radioactive tracer method. These authors employed the TiO_2 films prepared by thermal decomposition of titanium chloride (so, in principle, similar to those used in above mentioned XPS studies[102]).

Direct measurements of adsorption densities of the labelled anions, performed in dilute aqueous solutions of the corresponding acids, enabled to establish the following order of preferential adsorption: $H_2PO_4^- > HSO_4^- > Cl^- > ClO_4^-$. This sequence, based upon the results of co-adsorption experiments, refers to equilibrium adsorption conditions in the range of pH lower than the PZC of TiO_2.

The behaviour of the $H_2PO_4^-$ ions towards the TiO_2 surface differred markedly from that of the other ions examined by Kazarinov et al.[117]. Unlike, e.g., the sulphate ions, the phosphate ions exhibited still significant extent of adsorption in neutral and alkaline solutions, and underwent only slight desorption in pure water. All these observations confirm clearly the irreversible character of the phosphate adsorption.

3.3 Influence of Anions upon the Cathodic Behaviour of TiO_2

The kind of effects the anions may have upon the electrochemical behaviour of TiO_2 is illustrated by a series of cyclic voltammograms in Fig. 2, representing the cathodic reduction of $Fe(CN)_6^{3-}$ into $Fe(CN)_6^{4-}$ ions. These measurements were repeated in solutions containing three different supporting electrolytes, using titanium electrodes coated with a 10–15 μm thick TiO_2 film, identical to those employed in the XPS studies[103]. These electrodes exhibit the behaviour characteristic of a rather strongly doped n-type semiconductor, enabling to attain reasonably high cathodic currents. The main features of the

---

5 In fact, analogous observations have already been made for aluminium supporting thin, natural AlO(OH) film, shown to cause reduction of the NO_3^- and ClO_4^- ions into the $NH_3(NH_4^+)$, respectively, Cl^- species[115, 116].

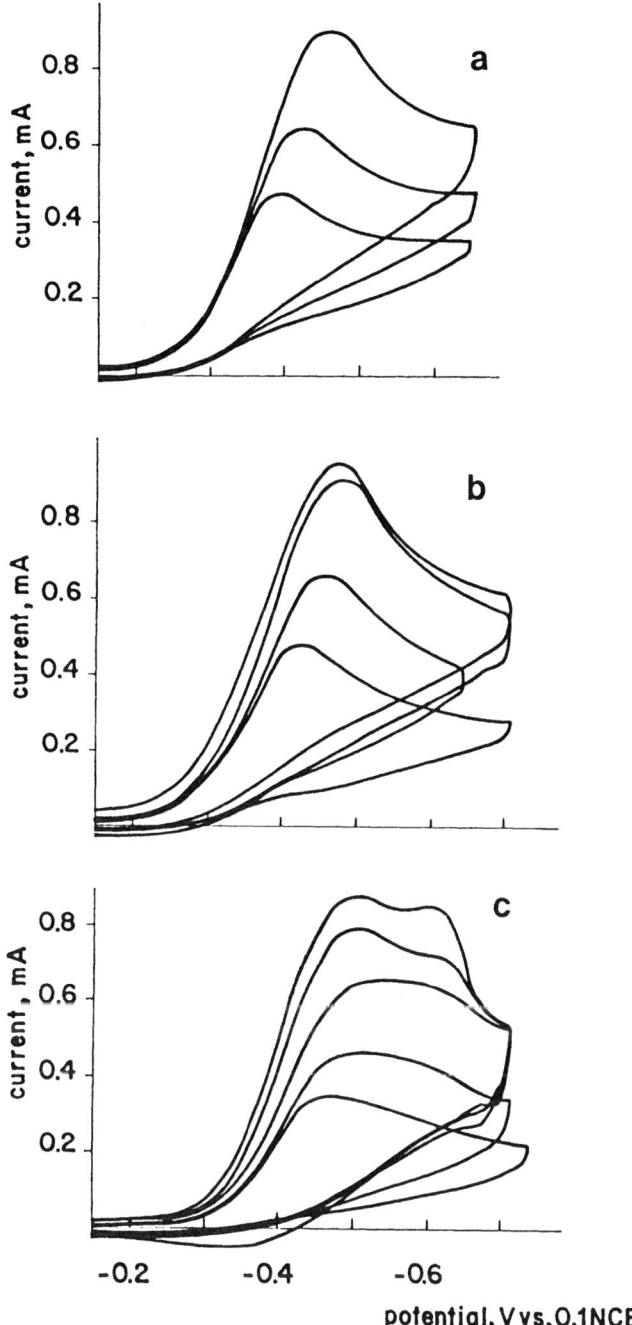

Fig. 2a–c. Cathodic cyclic voltammograms corresponding to the reduction of ferricyanide ions at a polycrystalline TiO_2 electrode (~ 0.28 cm^2). 0.01 M $K_3Fe(CN)_6$ solution with different supporting electrolytes: 0.5 M KNO_3 (part **a**); 0.5 M KCl (part **b**); 0.5 M K_2HPO_4/KH_2PO_4 (part **c**). Curves recorded (in order of increasing intensities) at 50, 100 and 200 mV · s^{-1}. *Two upper curves* in part **c** were obtained after a 15-h and a 60-h immersion of the electrode in the solution (200 mV · s^{-1})

voltammograms reproduced in Fig. 2, parts (a) and (b), are quite typical of an irrevers-ible charge transfer reaction. The only visible difference between the voltammetric pro-files, recorded in the solutions containing the KNO_3, respectively, the KCl supporting electrolyte, is the occurrence of the reduction maxima, in the presence of nitrates, at slightly less negative potentials than in the chloride solution. The replacement of the KNO_3 or KCl by the potassium phosphate supporting electrolyte brought about substan-tial changes both in the shape of the voltammograms and in the current intensities (cf. Fig. 2, part (c)). These changes are likely to be caused by the partial blocking of the surface sites by the adsorbed $H_2PO_4^-$ and/or HPO_4^{2-} anions.

Interestingly, a prolonged exposure of the titanium dioxide electrode to the ferricy-anide/phosphate solution resulted in a substantial increase of the observed reduction currents and in the appearance of a second cathodic peak (cf. Fig. 2, part c). The latter peak, shifted to more negative potentials, is presumably associated with the reduction of adsorbed $Fe(CN)_6^{3-}$ ions, occurring in addition to the diffusion-controlled reactions. These changes in the voltammograms are visibly related to slow adsorption processes at the TiO_2-solution interface. However, the exact way by which the $H_2PO_4^-(HPO_4^{2-})$ ions may affect the co-adsorption of the $Fe(CN)_6^{3-}$ ions remains unclear.

The above described voltammograms, relevant to the reduction of the ferricyanide ions, are a good illustration of the principal features of electrochemical behaviour of TiO_2. This behaviour is largely dependent on the relative positions of the equilibrium potential of the redox system present in the solution, on the one hand, and of the flat-band potential of the semiconductor on the other[118]. A rough estimate indicates that the potential of the $Fe(CN)_6^{3-}/Fe(CN)_6^{4-}$ couple is ca. 0.7–0.8 V more positive than the flat-band potential, V_{fb}, of TiO_2 in contact with the solution of pH 6–7 used for the experi-ments illustrated in Fig. 2. In such a case, a moderately doped, unilluminated titanium dioxide electrode is expected to act as a relatively efficient cathode but to block the passage of any significant anodic current. For this reason, no anodic wave, corresponding to the re-oxidation of the $Fe(CN)_6^{4-}$ ions, could be observed, even when the reverse sweep was prolonged up to high positive potentials. This does not preclude, however, that under different circumstances, in the presence of strongly reducing species with standard redox potentials more negative than the V_{fb}, the TiO_2 electrode may give rise to quite significant "dark" anodic currents. Such currents, associated with the electron injection by the reducing species into the conduction band of TiO_2, have, for example, been observed in the solutions containing Ti^{3+} or V^{2+} ions[119].

Typically, the rate of simple (outer-sphere) electron-transfer reactions, such as $Fe(CN)_6^{3-} + e^- \rightarrow Fe(CN)_6^{4-}$, is much slower at titanium dioxide than at metallic elec-trodes[6]. This is consistent both with the flat shape of the voltammetric peaks in Fig. 2 and their shift to more negative potentials with increasing the sweep rate. The kinetics of the cathodic reactions at TiO_2 appear to be markedly affected by the co-adsorption of some anions from the supporting electrolyte. The phosphate ions are not unique to cause such effects; arsenate, fluoride and certainly other anions are expected to act in a similar way.

---

6 Such behaviour is quite common to the semiconductors and also to the oxide-covered metal (especially valve metal) electrodes[118, 120]. It is not unusual that the rates of electron-transfer reactions on the latter materials are by 10 orders of magnitude smaller than on noble metal electrodes.

The observations of Augustynski et al.[103], regarding the reduction of the $Fe(CN)_6^{3-}$ ions in the solutions containing the KCl or KNO_3 supporting electrolyte are consistent with the direct electron transfer mechanism – without the mediation of surface states. This is supported both by the large values of the cathodic currents and their first-order dependence on the ferricyanide concentration in the solution. The direct electron transfer from the conduction band of TiO_2 to the $Fe(CN)_6^{3-}$ ions is expected to be favoured by the actual position of the $Fe(CN)_6^{3-}/Fe(CN)_6^{4-}$ redox potential with respect to the conduction band edge (cf. Fig. 3). In fact, the latter potential difference appears to be close to the value of the reorganization energy, $\lambda = 0.75$ eV, for the couple ferricyanide/ferrocyanide, evaluated by Vanden Berghe et al.[121].

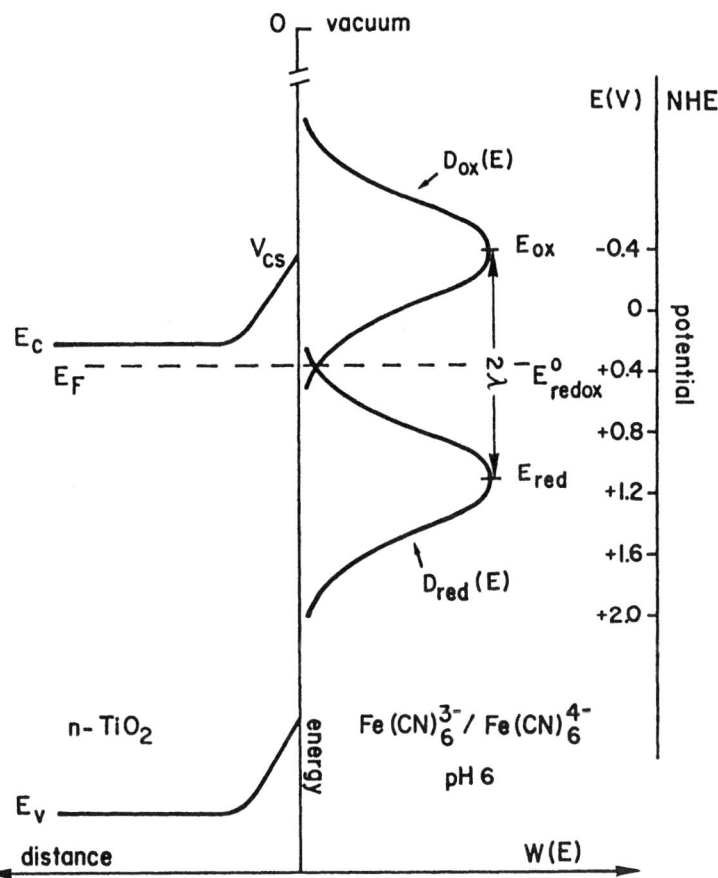

Fig. 3. Schematic energy diagram for titanium dioxide in contact with a $Fe(CN)_6^{3-}/Fe(CN)_6^{4-}$ solution. The redox couple is assumed to be present under standard conditions

4 Behaviour of Surface Peroxo Species in the Photoreactions at TiO_2

Extensive investigations of the photoelectrochemical cells, commenced in the early 1970s, have successively been extended to the "integrated" systems, employing semiconductors in a particle form suspended in an aqueous solutions[15-19, 21, 122-131]. Such semiconductor particles, most frequently containing small deposits of noble metals (e.g., platinum, rhodium) and, eventually, of noble metal oxides (e.g., ruthenium or iridium dioxide), when subjected to the band-gap illumination behave as short-circuited microelectrodes. The noble metal or noble-metal oxide deposits play in these systems a double role, enhancing the rate of certain reactions, such as, e.g., the hydrogen evolution (for Pt and Rh) and favouring the charge separation in the light absorbing particles. This second function becomes particularly important in the case of very small (e.g., colloidal) semiconductor particles[131, 132].

The microheterogeneous systems of such a kind permit to carry out endergonic processes in the same way as the photoelectrochemical cells employing macro-electrodes. However, the small spatial separation between anodic and cathodic sites in the former systems renders difficult the separation of products and, at the same time, increases the probability of back reaction.

Because of its satisfactory resistance towards photocorrosion and relatively high energy of the conduction band level, titanium dioxide (especially in its anatase form) has, together with strontium titanate, been widely used as a basis of the microheterogeneous systems. In particular, TiO_2 particles have been employed in a large number of experiments attempting to cleave water photochemically[15, 18-20, 125-128]. Most of these experiments have led to an intriguing observation that in the close systems, under conditions of room temperature and atmospheric pressure the hydrogen formation was not accompanied by the evolution of the stoichiometric amount of oxygen[19, 20, 133].

Detailed studies have shown peroxo species, present at the surface of TiO_2 particles and also in the solution, to be in this case the main product of the photo-oxidation of water[20, 133-135].

The difficulty, inherent to the experiments with dispersed TiO_2 particles, is that they do not allow to distinguish between the two possible routes by which the peroxo species can be formed at the surface of TiO_2, i.e.: (i) the photo-uptake of photogenerated molecular oxygen, involving electrons originating from the conduction band of TiO_2, which may, for example, be formulated as[7]:

$$O_{2(ad)} + 2\,H^+_{aq} + 2\,e^-(TiO_2) \rightarrow H_2O_2(TiO_2) \qquad (9)$$

or (ii) the photo-oxidation of water molecules by the positive holes created in the valence band of TiO_2:

$$2\,H_2O + 2\,h^+(TiO_2) \rightarrow H_2O_2(TiO_2) + 2\,H^+_{aq} \qquad (10)$$

---

7 This reaction is though to proceed through the formation of $O^-_{2(ad)}$ and/or $HO^-_{2(ad)}$ intermediates, followed either by their dismutation or further reduction[133, 136].
 The formulation of reaction (9) follows that given in the literature[135], $H_2O_2(TiO_2)$ being a general term denoting peroxide on the TiO_2 surface or in the solution. It is, however, important to note that the exact nature of the formed species is unknown.

To overcome this difficulty, Ulmann, de Tacconi and Augustynski[137] have employed a TiO$_2$ photoelectrode with the intent to compare the species photogenerated at open circuit (i.e., under conditions analogous to those of the experiments with titania suspensions) with the products formed at the photoelectrode subjected to an anodic bias. In particular, applying a sufficiently positive potential to the TiO$_2$ photoelectrode renders any back reaction involving conduction-band electrons (consequently, also reaction (9)) practically impossible, because of the negligibly small electron density on the surface of the semiconductor electrode[8].

The use of the electrode, for modelling the photoreactions occurring in the suspension of TiO$_2$ particles, enables, on the one hand, to follow the changes of photopotential at open circuit and to impose a chosen potential through an external bias, and, on the other hand, to monitor the species photogenerated at the surface of TiO$_2$ by means of linear-sweep voltammetry. The reduction peak arising when, after the photoexcitation of a TiO$_2$ electrode under anodic bias, the light is cut off and the potential is swept in the cathodic direction, has originally been assigned by Wilson[138] to the surface states, supposedly associated with an intermediate of the oxygen evolution.

In this connection, it is to be pointed out that a proper detection of the products of the photoanodic reaction requires the corresponding experiment to be performed in thoroughly deaerated solution. In fact, the reduction of the surface-bonded species, formed at the TiO$_2$ photoanode, is observed in the region of cathodic potentials close to that in which the reduction of dissolved oxygen takes place. This coincidence has probably been the cause of recent contradictory interpretations of voltammetric measurements effected with TiO$_2$ electrodes. Thus, the excess cathodic current, resulting from the pre-exposure of a TiO$_2$ electrode, in an oxygen-saturated 1 M KOH solution, to the band-gap illumination under anodic bias, has been explained in terms of a local supersaturation of the solution by the photogenerated oxygen (Salvador and Gutiérrez[139]). On the other hand, in the case of a similar experiment carried out in an acidified 1 M Na$_2$SO$_4$ solution, the voltammetric reduction peak, observed at potentials slightly more positive than the flat-band potential of TiO$_2$, has been assigned to photogenerated hydrogen peroxide species chemisorbed on the surface of TiO$_2$ (Salvador and Gutiérrez[140]). The latter authors[139-140] have not specified if the suggested difference in the nature of the intermediates formed at the TiO$_2$ photoanode in alkaline, respectively, in acidic solution is to be associated with a change in the mechanism of oxygen photogeneration.

Subsequently, Ulmann et al.[137] have established that relatively stable (i.e., persistent several hours after the electrode illumination had been stopped), surface-bonded species are photogenerated at TiO$_2$ in alkaline solutions, both at open circuit and under anodic bias. The main scope of the latter work was to characterize electrochemically the species accumulating on the surface of titanium dioxide exposed to the near-UV illumination ($\lambda > 335$ nm), in a sodium hydroxide solution, under conditions simulating: a) photochemical cleavage of water; b) oxygen photo-uptake; and c) photoassisted anodic (under external bias) oxidation of water.

---

8 The density of electrons on the surface of an n-type semiconductor, n_s, changes with the applied potential, E_{app}, as[138]

$$n_s = n_b \exp[- e(E_{app} - V_{fb})/kT] \qquad (11)$$

where n_b is the bulk electron density of the semiconductor, V_{fb} is its flat-band potential, e is the elementary charge, k is Boltzmann constant, and T is absolute temperature.

4.1 Electrochemical Characterization of the Species Photogenerated at the Surface of TiO_2

In Fig. 4 (extracted from the work of Ulmann et al.) are represented cyclic voltammograms recorded in 0.1 M deaerated aqueous NaOH after the TiO_2 electrode had been illuminated at a potential of 0.57 V versus reversible hydrogen electrode in the same solution, RHE, (the photoanodic sequence)[9] and, subsequently, its potential was swept in the cathodic direction: (i) within 1 s after the cut-off of illumination (for curve a) or (ii) after additional 2 min in the dark, during which for 1 min the solution had been vigorously stirred with argon (for curve c). A voltammogram practically identical with curve c was also obtained when the solution had additionally been stirred during the entire photoanodic sequence. Thus, the amount of charge accounting for the difference between voltammograms a and c in Fig. 4 is to be associated with the reduction of photogenerated oxygen and, possibly, of other dissolved species formed at the photoanode. Importantly, the decrease of the cathodic peak, occurring during the first 2 min after the light-off, was less than 20 per cent of the intensity of peak a in Fig. 4. Further decay of the cathodic peak was distinctly slower as shown by the peak current in curve e (Fig. 4), recorded 16 1/2 h after the light-off, which still exceeded 50 per cent of the

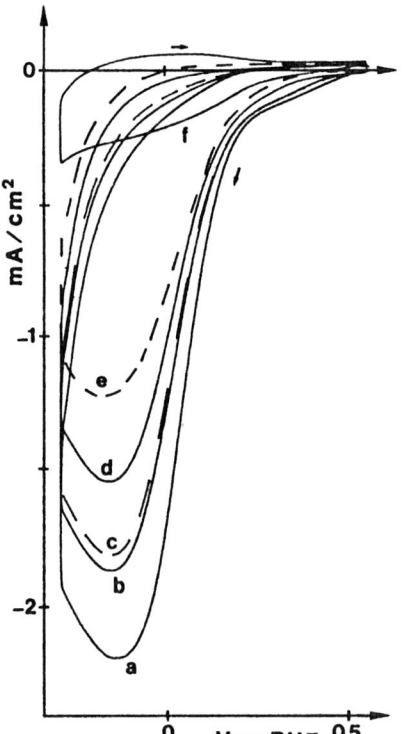

Fig. 4. Cyclic voltammograms showing reduction (in the dark) of the species photogenerated at the TiO_2 electrode polarized anodically for 2 min in deaerated 0.1 M aqueous NaOH at 0.57 V; $\lambda > 335$ nm. Cathodic sweep of the potential, at 50 mV $\cdot$ s^{-1}, started within 1 s after the light-off (curve a); after additional 2 min (curves b and c); after 1 h (curve d); and after 16 1/2 h (curve e). Solution was stirred with Ar during the period in the dark preceding recording of curves c, d, and e. Curve f is representative of the second and following sweeps[137]

---

9 The corresponding photocurrent was quite reproducible, decreasing from the initial value of ~ 400 μA to ~ 370 μA.

Table 2. Mean current efficiencies for the photogeneration of the surface-bonded species at the TiO$_2$ electrode polarized anodically for increasing periods of time

Duration of the photo-anodic sequence, s	Anodic photocurrent at 0.57 V[a], mA	Cathodic charge associated with the reduction of surface-bonded species[b], mC	Cathodic/anodic current efficiency
30	0.46/0.43	2.9	~22%
60	0.45/0.42	3.35	~13%
120	0.45/0.41	4.0	~ 8%
300	0.45/0.41	4.7	~ 4%
600	0.43/0.38	–	–

[a] Initial and final photocurrents observed during photoanodic sequence ($\lambda > 335$ nm)
[b] Obtained by integration of the cathodic voltammetric curves in Fig. 4 and subsequent subtraction of the background charge (from the second and following sweeps in the dark)

initial peak current (curve a). During all that period the electrode was maintained in darkness at 0.57 V (i.e., at the potential applied for the photoanodic sequence) and the solution was vigorously stirred with argon. Curve f in Fig. 4, corresponding to the second and subsequent sweeps in the dark, is characteristic of the behaviour of polycrystalline TiO$_2$ in solutions not containing other reducible species than water or H$_3$O$^+$ ions[142]. The magnitude of the cathodic current in voltammogram f has been interpreted in terms of the surface reduction of TiO$_2$[137].

$$ // (Ti^{4+})_4(O^{2-})_8 + nH_2O + ne^- \rightarrow (H^+)_n(Ti^{4+})_{4-n}(Ti^{3+})_n(O^{2-})_8 + nOH^- \tag{12} $$

where // symbolize vacant octahedral sites.

Ulmann et al. have also observed that, with increasing the duration of the photoanodic sequence, the intensity of the cathodic peak increased much slower than the magnitude of the corresponding photoanodic charge. Their results are summarized in Table 2 in terms of faradaic efficiency for the production of the species bound to the surface of the TiO$_2$ electrode[10]. While, for a relatively short (30 s) photoanodic sequence, more than 20 per cent of the consumed anodic charge was associated with the formation of surface-bonded species[11], this ratio dropped to about 4 per cent for a 5 min period of electrode illumination.

Changes in the voltammograms recorded by Ulmann et al., associated with an increase of the amount of the reducible species, (cf. Fig. 5) such as the broadening of the reduction peak and its shift to more cathodic potentials, were typical of an irreversible surface process. Such behaviour is probably due to the fact that, with increasing the coverage, the surface-bonded species continued to be formed inside the pores in the TiO$_2$ electrode (cf. Fig. 6). This would imply an increased ohmic resistance for the consecutive reduction process. An alternative explanation could involve a rearrangement ("aging")

---

10 The conditions of those experiments (i.e., the delay of 1 min between the light-off and the beginning of the cathodic sweep) were chosen to render negligible the contribution to the cathodic peak arising from the soluble photogenerated species.
11 Molecular oxygen was the principal photogenerated product; O$_2$ evolution was clearly visible during all the photoanodic experiments.

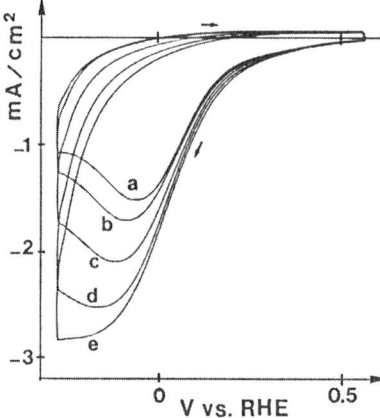

Fig. 5. A sequence of cathodic cyclic voltammograms, recorded at 50 mV · s^{-1} in the dark, for the TiO$_2$ electrode pre-exposed to the UV light (λ > 335 nm) at 0.57 V for 30 s; 60 s; 2 min; 5 min; and 10 min for, respectively, *curves a, b, c, d,* and *e.* Deaerated 0.1 M aqueous NaOH. After the light-off the solution was stirred with Ar for 60 s before the start of the cathodic sweep[137]

of the surface species, as observed, e.g., for formation and reduction of the surface oxides at noble metals[143].

Ulmann et al. have also shown that making more positive the potential, applied during the photoanodic sequence, resulted simply in an increase of the reduction peak accompanied by its shift to more negative potentials. No changes in the shape of the voltammograms, which would suggest the presence of different photogenerated species, were observed, even for the potential of the photoanodic sequence higher than the reversible potential of oxygen evolution in the dark equal to 1.229 V versus RHE[144].

All the measurements described above were performed in deaerated 0.1 M aqueous NaOH which enabled to minimize the contribution to the reduction current arising from dissolved photogenerated oxygen. In Fig. 7 are compared cathodic voltammograms

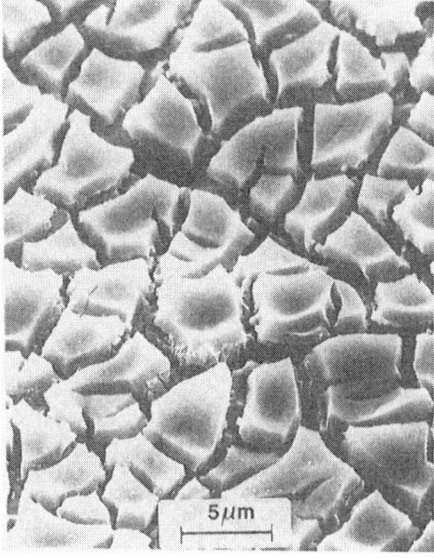

Fig. 6. Scanning electron micrograph showing microcracked surface of the TiO$_2$-film electrode[137]

recorded after a TiO_2 electrode had been illuminated, under identical anodic bias, respectively in deaerated (curve c) and O_2-saturated (curve d) 0.1 M aqueous NaOH[137]. The significant additional current in curve d corresponds to the reduction of dissolved molecular oxygen, occurring in the same potential region in which reduction of photo-generated, surface-bonded species takes place. However, the peak associated with the former reaction (curve b) may be easily recognized, its maximum being distinctly more anodic than that of the peak due to the surface process (cf. curve c in Fig. 7).

Exposure of a TiO_2 electrode to the band-gap illumination in a deaerated sodium hydroxide solution, in the absence of any external polarization, has been observed[137] to bring apparently about formation of the same species which were produced under anodic bias. In fact, both the shape and the potential of the cathodic peaks, visible in Fig. 8, corresponding to the reduction of the species accumulated during illumination of the TiO_2 electrode at open circuit, are quite similar to the features of the peaks in Fig. 4. However, the amount of cathodically reducible species, formed during a prolonged (14 h) irradiation of the electrode under open-circuit conditions (cf. curve d in Fig. 8), was relatively small – somewhat less than that resulting from a 30-s anodic polarization at 0.57 V (cf. Table 2).

It is useful to recall, in this connection, that, in open circuit, the photoanodic (valence-band) reaction, responsible for the formation of the surface-bonded species[12], may

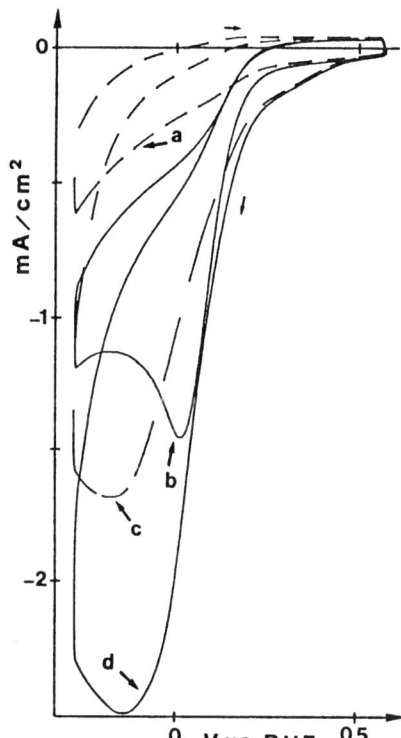

Fig. 7. Cyclic-voltammetric i-E profiles of the TiO_2 electrode, obtained in the dark in deaerated (for *curves a* and *c*) and in O_2-saturated (for *curves b* and *d*) 0.1 M aqueous NaOH. *Curves c* and *d* are for the electrode prepolarized for 2 min at 0.57 V under UV illumination ($\lambda > 335$ nm). Delay of 2 min between the light-off and starting of the cathodic sweep. S = 50 mV · s^{-1} [137]

---

12 Oxygen being (at least initially) absent from the solution, the cathodically-reducible, surface-bonded species may be considered as the net product of an anodic reaction – whatever would be the detailed mechanism of their formation.

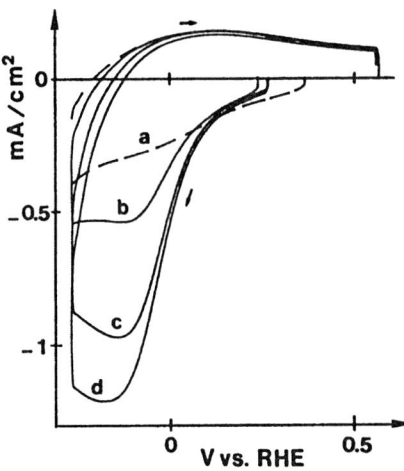

Fig. 8. Cyclic voltammograms, recorded in the dark at 50 mV · s⁻¹, in deaerated 0.1 M aqueous NaOH, illustrating cathodic reduction of the species photogenerated at the TiO_2 electrode in open circuit. Time of exposure to UV illumination ($\lambda > 335$ nm): 2 min; 30 min; and 14 h for, respectively, *curves b, c,* and *d.* Delay of 2 min between the light-off and starting of the cathodic sweep. *Curve a* is representative of the second and following sweeps[137]

proceed only in the presence of an associated cathodic (conduction-band) reaction draining the photogenerated electrons. The open-circuit potential (photopotential), equal to 0.13 V versus, RHE, observed under conditions of the above experiment[137], rules out hydrogen evolution as a feasible cathodic reaction. The alternative possibility is the surface reduction of TiO_2, according to reaction (12). This implies that, in the absence of other reducible species, the extent of reduction of the TiO_2 surface should be equal to the coverage by the surface-bonded species.

The injection of oxygen to the solution resulted in an increase in the rate of formation of photogenerated surface species at open circuit[137]. This may be explained by the fact that, in the case of O_2 photo-uptake, the surface-bonded species are partly formed *via* cathodic reduction of dissolved molecular oxygen. As a matter of fact, parallel experiments, consisting in a controlled-potential reduction of dissolved O_2 at the TiO_2 electrode in the dark, revealed the presence of the surface species similar to those generated during illumination of the electrode[13].

All the results discussed above strongly suggest that the exposure of the TiO_2 (anatase) electrode immersed in an alkaline solution to the band-gap illumination, under anodic bias or at open circuit, leads in all cases to the formation of very similar (if not identical) species bound to the oxide surface. Such conclusion is supported by the similarity of the potential, shape and decay characteristics of the corresponding cathodic reduction peaks shown in Figs. 4, 5, 7 and 8. These species have been identified, in the course of earlier experiments performed with aqueous alkaline dispersions of TiO_2 (anatase) particles[14] irradiated with near-UV light, as surface-bonded, peroxo-titanium complexes[133, 135].

---

13 This experiment consisted in polarizing the TiO_2 electrode in the dark at a constant potential of 0.17 V, corresponding to the photopotential measured during the photo-uptake of oxygen. The amount of the surface species formed *via* cathodic reduction of O_2 in the dark was less than 30 per cent of the amount photogenerated at the TiO_2 electrode irradiated ($\lambda > 335$ nm) at open circuit for the same period of time.

14 It is assumed here that a large surface TiO_2 photoelectrode mimics closely the behaviour of the TiO_2 particles of not too small size, and, in particular, that the surface species, formed during irradiation of the TiO_2 dispersion, respectively, the TiO_2 electrode at open circuit, are identical.

In order to check the stability of the surface peroxo species, Ulmann et al.[137] have examined the decay of the cathodic peak in the presence of light of various wavelengths. To this end, the electrode had first been polarized anodically for 2 min under standard conditions (E = 0.57 V; λ > 335 nm) and then, after the potentiostat was switched off and the filter changed, was irradiated for 10 min at open circuit before starting the cathodic sweep. Results presented in Fig. 9, showing a marked decrease of the reduction peak with increasing the portion of the UV light, apparently suggest that the accumulated surface peroxo species undergo partial photodecomposition. However, one cannot neglect that the change of wavelength and intensity of the incident light leads to a modification of the photopotential. When, instead of open-circuit exposure to light of various wavelengths, the TiO_2 electrode was polarized in the dark for 10 min at a potential equal to the corresponding photopotential, the decay of the cathodic peak was quite similar to that shown in Fig. 9.

In the light of the latter experiment, the surface peroxo species appear as a preferred target for the conduction-band electrons and, consequently, as particularly efficient recombination centers. In this connection, it is to be noted that all the cathodic voltammograms, representing the reduction of the surface-bonded species photogenerated under anodic bias, exhibited significant currents already at potentials of 0.2–0.4 V versus RHE, i.e., distinctly more positive than the flat-band potential of TiO_2.

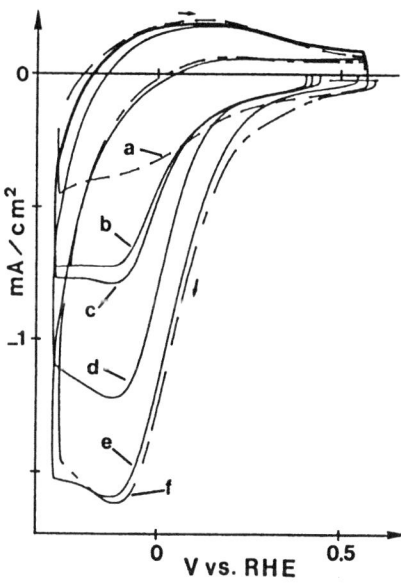

Fig. 9. Effect of the spectral composition of the incident light on the decay of the cathodic voltammetric peak associated with the reduction of surface-bonded species photogenerated at the TiO_2 electrode polarized for 2 min at 0.57 V under UV illumination (λ > 335 nm). The electrode was subsequently illuminated for 10 min at open circuit with the light of λ > 305 nm *(curve b);* λ > 335 nm *(curve c);* λ > 360 nm *(curve d);* λ > 400 nm *(curve e);* and kept in the dark *(curve f).* *Curve a* is representative of the second and following sweeps. All cyclic voltammograms were recorded in the dark at 50 mV · s^{-1} in deaerated 0.1 M aqueous NaOH[137]

4.2 Reduction Behaviour of Hydrogen Peroxide
    at the TiO_2 Electrode

In view of its possible role of intermediate in the formation of the surface peroxo-titanium species, Ulmann et al.[137] have also examined the cathodic reduction of hydrogen peroxide in NaOH solution. Figure 10 shows a typical, narrow-shaped cathodic voltammogram, recorded for the TiO_2 electrode in 0.1 M NaOH/10^{-3} M H_2O_2 solution, compared to that corresponding to the reduction of photogenerated peroxo-titanium species (dotted line curve). The difference between the two peak potentials, E_p, amounts to ca. 0.3 V, with the value of E_p, characteristic of the reduction of H_2O_2, being more positive than the flat-band potential of TiO_2. It is important to note here that the peak associated with the reduction of hydrogen peroxide is still shifted of ca. 0.15 V to more positive potentials with respect to the voltammetric peak corresponding to the reduction of dissolved molecular oxygen at the TiO_2 electrode (cf. curve b in Fig. 7).

The above voltammetric measurements were also extended to the TiO_2 electrodes pretreated in H_2O_2 solution in water, before being immersed in 0.1 M aqueous NaOH (Fig. 11). In fact, hydrogen peroxide is known to undergo chemisorption on titanium dioxide with the limiting coverage attaining (in 0.08 M H_2O_2 solution) ca. $1.8 \cdot 10^{14}$ molecules per real cm^2 of the anatase surface[145]. The voltammograms obtained for the reduction of H_2O_2 preadsorbed at the TiO_2 electrode were quite similar to those recorded during analogous measurements performed in the presence of hydrogen peroxide in the 0.1 M NaOH solution. Indeed, the peak associated with the reduction of H_2O_2 preadsorbed from its 10^{-3} M aqueous solution, shown in Fig. 11, is only slightly more cathodic ($\Delta E_p \simeq 0.04$ V) and, as expected, less intense than the corresponding peak in Fig. 10.

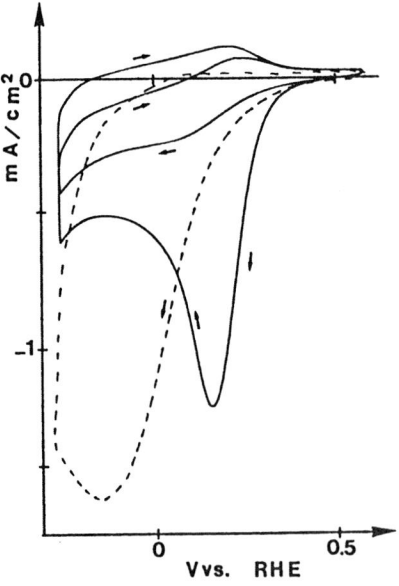

Fig. 10. Cyclic-voltammetric i-E profile for the cathodic reduction of H_2O_2 at the TiO_2 electrode. *Full-line curve* recorded in 0.1 M aqueous NaOH/ 10^{-3} M H_2O_2. Other conditions as in Fig. 11

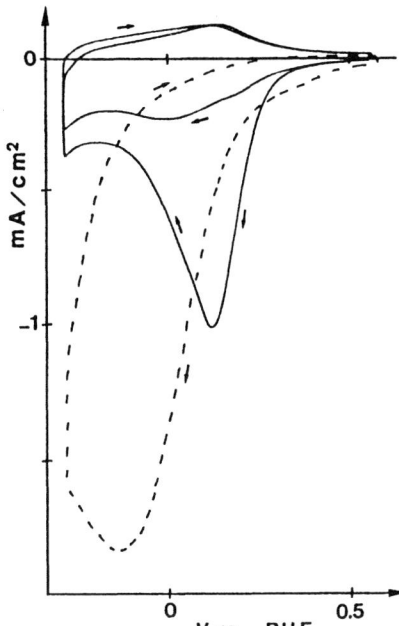

Fig. 11. Cyclic-voltammetric i-E profile for the cathodic reduction of H_2O_2 at the TiO_2 electrode. *Full-line curve* recorded in 0.1 M NaOH for the electrode being immersed for 2 min in 10^{-3} M aqueous H_2O_2. *Dotted-line curve* corresponds to the reduction of the peroxo-titanium species photogenerated at the TiO_2 polarized for 2 min at 0. 57 V ($\lambda > 335$ nm). Voltammograms were recorded in the dark at 50 mV $\cdot$ s^{-1} [137]

4.3 Possible Intermediates of the Photo-Oxidation Reactions at TiO_2

The results of the electrochemical measurements discussed above provide convincing evidence that the surface species, responsible for the "oxygen storage" at irradiated TiO_2 particles, identified as peroxo-titanium complexes[20, 133-135], are also produced on a TiO_2 photoelectrode subjected to moderate or high anodic bias. Now, the large band bending, occurring under the latter conditions[137], excludes practically that a simultaneous cathodic reaction, such as the oxygen reduction (cf. Eq. (9)), might concur to the formation of the surface peroxo species. The reason for this is the exponential decrease of the surface electron density with increasing the anodic bias, according to Eq. (11), rendering any partial cathodic current negligible for the band bending higher than ca. 0.5 V[137]. Moreover, if the surface peroxo species were formed *via* the cathodic reduction of photogenerated molecular oxygen, such reaction would hardly be effective at potentials more positive than the standard redox potential of the O_2/HO_2^- couple:

$$O_2 + H_2O + 2\,e^- \;\rightarrow\; HO_2^- + OH^- \tag{13}$$

equal to 0.752 V versus RHE (at pH 14)[144]. Now, the photogeneration of the peroxo species has been demonstrated to occur at the TiO_2 electrode polarized anodically at a potential as high as 1.47 V[137]. Hence, the only conceivable way, by which formation of the surface peroxo species may take place under large positive bias, is an anodic process involving photogenerated positive holes from the valence band.

The latter conclusion, obviously, does not preclude that, under conditions corresponding to the photodecomposition of water over titania suspensions (imitated by a TiO_2 photoelectrode left at open circuit), the surface peroxo species may also be formed through a cathodic route (see, for example, Eq. (9)). This should, in particular, be true when the photolysed solution initially contains dissolved molecular oxygen. In fact, experiments performed by Ulmann et al.[137], consisting in a controlled-potential reduction of oxygen at the TiO_2 electrode (in the dark) and the subsequent voltammetric characterization of the products of this reaction, provide direct evidence for the formation of the surface species similar but not necessarily identical with those photogenerated under significant anodic bias.

4.3.1 The Role of Hydrogen Peroxide

Hydrogen peroxide has frequently been invoked as a possible intermediate and co-product of the photo-oxidation of water at TiO_2[140, 146–149] although no direct experimental evidence supporting these claims has been provided.

In this respect, the behaviour of another semiconducting oxide with a large band gap, ZnO, appears much less controversial. In fact, it is well established since a long time that the oxygen-containing aqueous suspensions of zinc oxide, irradiated with UV light, produce easily detectable amounts of H_2O_2[150–158]. The concentrations of photogenerated hydrogen peroxide as high as 10^{-3} mol $\cdot$ dm^{-3} have been observed in the solutions containing certain organic additives such as acids, alcohols, amides, etc. In the ZnO/O_2 aqueous systems without additive, the quantum yield of peroxide formation decreases to zero as the H_2O_2 concentration reaches about 10^{-5} mol $\cdot$ dm^{-3}[154].

The use of ^{18}O – enriched gaseous oxygen allowed to demonstrate that the photogenerated hydrogen peroxide originated entirely from O_2[155]. In other words, the crucial step in the reaction sequence responsible for the formation of H_2O_2 consists in the electron transfer from the conduction band of ZnO to the adsorbed molecular oxygen. For instance, Calvert et al.[155] have proposed the following pathway:

$$O_2 + e^- \rightarrow O_2^- \tag{14}$$

$$O_2^- + H_2O \rightleftarrows HO_2 + OH^- \tag{15}$$

$$2 HO_2 \rightarrow H_2O_2 + O_2 \tag{16}$$

The electrons injected to the conduction band of zinc oxide by some strongly reducing agents may here play the same role as photogenerated electrons, leading to the formation of hydrogen peroxide in the dark[156, 159].

In contrast with the behaviour of zinc oxide, the question of the eventual photogeneration of H_2O_2 at titanium oxide remains controversial. Early comparative experiments performed with irradiated aqueous suspensions of ZnO and TiO_2[153] allowed to prove the formation of hydrogen peroxide at the former but not at the latter semiconductor. However, more recently, Pappas and Fischer[146], using a sensitive enzymatic-spectroscopic method, were able to detect hydrogen peroxide in O_2-saturated aqueous suspensions of two (among three tested by them) commercial TiO_2 pigments. The amount of H_2O_2 present in the solution was much larger for the anatase than for the rutile sample and followed chalking tendencies of the corresponding pigments.

Photosynthetic production of H_2O_2 and H_2 on UV-irradiated suspensions of titanium dioxide in deionized water has been claimed by Rao et al.[147]. Their experiments were performed with strongly reduced TiO_2 powder and the reaction mixture was thoroughly deaerated. Unfortunately, the procedure chosen by Rao et al. for analysing hydrogen peroxide, consisting in titrating the entire reaction mixture, did not allow to discriminate between the peroxides accumulated on the surface of the photocatalyst and those eventually released into the solution. Harbour et al.[160] have employed an in situ method to monitor UV-irradiated, O_2-containing, aqueous dispersion of titanium dioxide for the eventual presence of hydrogen peroxide. The method consisted in measuring, by means of a Clark electrode, the changes in oxygen concentration consecutive to the injection of catalase to the system. An evidence for the transitory generation of H_2O_2 (in the form of a slight increase of the oxygen concentration) could be obtained in the presence, but not in the absence, of sodium formate added to the irradiated TiO_2 dispersion. Importantly, no H_2O_2 was detected in the final reaction mixture, after the light had been switched off. This contrasts with the behaviour of such photocatalysts as zinc oxide or cadmium sulfide for which the addition of catalase was effective in detecting H_2O_2 accumulating in the solution during the photo-uptake of oxygen[158, 161]. It is symptomatic in this regard that, unlike TiO_2, both ZnO and CdS tend to photo-corrode when exposed to water in the presence of UV light.

Hydrogen peroxide has also been detected after prolonged operation of water photo-electrolysis cells including a TiO_2 photoanode and a Pt cathode[162, 163]. The H_2O_2 formation at the photoanode has been proposed to account for the observed volume ratio of the evolved oxygen to hydrogen lower than $1:2$[162, 163]. However, this explanation appears rather unlikely, as the oxygen imbalance has only been observed in an one-compartment photoelectrolysis cell, whilst in a two-compartment cell, operating under similar conditions, the two gases were produced in a stoichiometric ratio[162]. This indicates that hydrogen peroxide must be formed by the reduction of photogenerated oxygen at the Pt cathode rather than by the direct photo-oxidation of water at TiO_2.

It is important to note in this connection that experiments, undertaken with a rotating Pt ring-TiO_2 disk electrode (RRDE), failed to demonstrate the photogeneration of H_2O_2 at anodically polarized, illuminated TiO_2 disk[149]. Still, the RRDE technique has been successful in detecting H_2O_2 produced through the cathodic reduction of dissolved oxygen at an unilluminated TiO_2 disk electrode[164, 165]. The latter studies have shown that the mechanism of the oxygen reduction at titanium dioxide is strongly affected by the pH of the solution. Thus, in alkaline solution (1 M aqueous KOH), according to Parkinson et al.[164], the reduction reaction follows essentially a 4-electron pathway, leading to the formation of OH^- ions

$$O_2 + 2\,H_2O + 4\,e^- \rightarrow 4\,OH^- \tag{17}$$

with hydrogen peroxide being a side product accounting for about 5 per cent of the total cathodic current. This observation is consistent with the results of voltammetric measurements, performed in 0.1 M aqueous NaOH, showing (Figs. 7 and 10) that the reduction of HO_2^- ions at a TiO_2 electrode takes place at potentials distinctly more positive than that of dissolved molecular oxygen[137]. In contrast with the behaviour observed in alkaline solutions, hydrogen peroxide appears to be the main product of the oxygen reduction at the TiO_2 cathode in acidic media. In fact, experiments performed with the

RRDE arrangement in 0.5 M aqueous H_2SO_4 by Kobayashi et al.[165] have demonstrated current efficiencies for the formation of H_2O_2 at TiO_2

$$O_2 + 2\,H_{aq}^+ + 2\,e^- \rightarrow H_2O_2 \tag{18}$$
(overall reaction)

as high as 70 per cent. Importantly, for the platinised TiO_2 electrode, the observed ratio of H_2O_2 decreased to about 15 per cent[165]. It is to be noted, however, that the direct comparison of the latter results is rendered slightly problematic by the fact that the potential applied to the platinised TiO_2-disk electrode was at least 0.3 V more positive than that imposed to the TiO_2 electrode without platinum. The latter potential, ca. -0.05 V versus RHE, was close to the flat-band potential of TiO_2 and, consequently, did not reflect properly the conditions occurring during the open-circuit photo-uptake of oxygen.

When considering the behaviour of platinised TiO_2 one has to take into account that the mechanism and products of the cathodic reduction of oxygen at a platinum electrode are potential-dependent[166, 167]. In particular, the potentials lower than ca. 0.2 V versus RHE[15], at which platinum becomes covered with adsorbed hydrogen, are known to favour the formation of H_2O_2[166]. To account for such behaviour, the possibility of a direct chemical reaction between the molecular oxygen and adsorbed hydrogen

$$O_2 + 2\,H_{ad} \rightarrow H_2O_2 \tag{19}$$

has been invoked[166]. This reaction may, under certain conditions, play an active role in the process of the O_2 photo-uptake at platinised TiO_2 suspensions. This must be true especially for Pt-loaded anatase suspensions which, when irradiated with an intense UV light, should attain a sufficiently low potential to enable the occurrence of reaction (19).

The fact that the photosynthesis of hydrogen peroxide on irradiated titanium dioxide suspensions is manifestly ineffective is easily understood when one examines together the cathodic and the photoanodic behaviour of H_2O_2 at a TiO_2 electrode[16]. Thus, the reduction of the HO_2^- ions

$$HO_2^- + H_2O + 2\,e^- \rightarrow 3\,OH^- \tag{20}$$

starts already at potentials significantly (ca. 0.3 V) more positive than that of the RHE (cf. Fig. 10). This implies, among others, that no hydrogen evolution

$$2\,H_2O + 2\,e^- \rightarrow H_2 + 2\,OH^- \tag{21}$$

should be expected, even at the noble-metal-loaded TiO_2, until the concentration of H_2O_2 reaches a negligibly small level.

Apart from the easy cathodic reduction of the HO_2^- ions, also their photo-oxidation at the TiO_2 electrode proceeds apparently at a high rate. This is illustrated by the fact that

---

15 The amount of H_2O_2 formed during oxygen reduction at the Pt/TiO_2 electrode has been determined by Kobayashi et al.[165] for a (disk) potential higher than 0.2 V vs RHE.

16 The following analysis pertains strictly to the behaviour of TiO_2 in alkaline solutions.

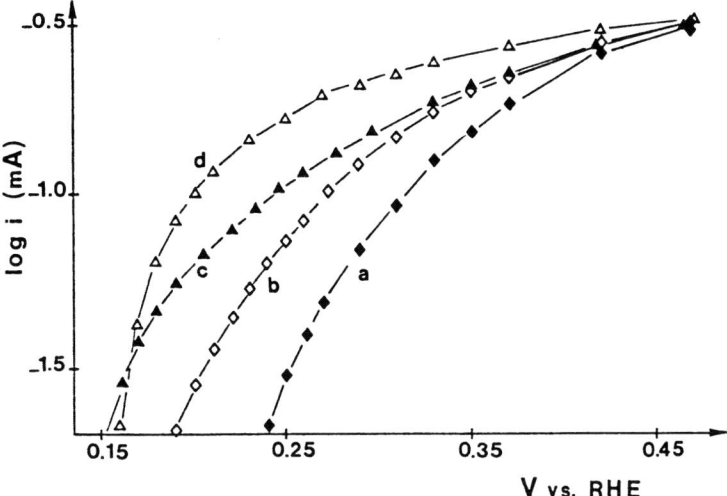

Fig. 12. Potentiodynamic (at $1\,mV \cdot s^{-1}$) i-E curves for the TiO_2 electrode illuminated ($\lambda > 335$ nm) in 0.1 M aqueous NaOH *(curve a);* 0.1 M NaOH/10^{-4} M H_2O_2 *(curve b);* 0.1 M NaOH/10^{-3} M H_2O_2 *(curve c)* and 0.1 M NaOH/10^{-2} M H_2O_2 *(curve d)*[137)]

an addition of H_2O_2 to NaOH solution causes a marked increase of anodic photocurrent in the initial part of the photocurrent-voltage curve (Fig. 12)[137)]. Since the amount of stationary photocurrent measured at low applied bias, is in principle determined by the competition between the charge-transfer reaction and surface recombination involving minority carriers[168)], the role of H_2O_2 should be to affect the rate of the former or of the latter process. As hydrogen peroxide seems very unlikely to block the recombination sites at the TiO_2 surface, the observed increase of the photocurrent must be the result of a large value of the rate constant for the charge-transfer reaction

$$HO_2^- + OH^- + 2\,h^+ \rightarrow O_2 + H_2O \tag{22}$$

The above results of photoelectrochemical measurements are entirely consistent with the observations reported by Brown and Darwent[169)] concerning competitive photo-oxidation of Methyl Orange (($CH_3)_2NC_6H_4N=NC_6H_4SO_3^- Na^+$) and hydrogen peroxide at colloidal TiO_2 suspensions. In the absence of H_2O_2 from the solution the photoreactions occuring at irradiated ($250 < \lambda < 380$ nm) TiO_2 consisted in the oxidation of Methyl Orange and simultaneous reduction of dissolved oxygen. Addition of H_2O_2 to the reaction mixture resulted in a pronounced inhibition of the Methyl Orange photo-oxidation replaced by that of the hydrogen peroxide. The kinetic analysis of the reaction effected by Brown and Darwent demonstrated that H_2O_2 reacts directly with a precursor to the species responsible for the dye oxidation, i.e., with the photogenerated positive holes, h^+. The rate constant for this reaction

$$H_2O_2 + 2\,h^+ \rightarrow O_2 + 2\,H_{aq}^+ \tag{23}$$

determined at pH 11.2 appears to be 1000 times larger than that for the formation of $Ti_s - O^{\cdot}$ radicals

$$Ti_s - O^- + h^+ \rightarrow Ti_s - O^{\cdot} \tag{24}$$

considered by the above authors as intermediates in the Methyl Orange oxidation. Accordingly, in the absence of hydrogen peroxide, the electron-hole recombination results to be the main process, only one among every 450 photogenerated holes leading to the formation of the $Ti_s - O^{\cdot}$ radicals. In contrast, H_2O_2 at a concentration of 10^{-4} mol · dm^{-3} was shown to intercept about 70 per cent of all photogenerated holes before they recombine with the electrons.

A feature of Fig. 12 which deserves further comment is that the addition of hydrogen peroxide to 0.1 M aqueous NaOH leads not only to the increase of the photocurrent but also to the shift of its apparent onset to less positive potentials[17]. Such behaviour seems to be quite typical for the ions interacting strongly with the semiconductor surface (for example I^- ions adsorbed at the WSe_2 electrode)[168, 171]. The interaction of this kind is believed to produce an additional potential drop in the Helmholtz part of the double layer, leading to the shift of the flat-band potential to more negative values[168]. It is important, however, to note in this connection that a similar significant shift of the photocurrent onset potential at TiO_2 may also be induced by the presence of some uncharged species (for example, by the addition of methanol to a 0.1 M NaOH solution – as will be further discussed in Sect. 5).

The photodecomposition of hydrogen peroxide, occurring at the TiO_2 suspension or the TiO_2 electrode at open circuit, may be visualized as a result of action of the local cells involving reactions (20) and (22) as the cathodic, respectively, photoanodic process. Considering the observed photopotential as a mixed potential, an estimate of the local-cell current is provided by the value of the cathodic (dark) current at a potential equal to the photopotential (cf. Fig. 13). By subtracting the (negative) cathodic (dark) current, i_{cath}, from the experimentally determined photocurrent, i_{ph}, one obtains for each potential the effective photoanodic current, $i_{ph.an} = i_{ph} - i_{cath}$ (curve c in Fig. 13). The absolute values of the cathodic and photoanodic currents at the photopotential, $|i_{cath}| = i_{ph.an}$, represent the rate of photoelectrochemical decomposition of the HO_2^- ions for a given concentration of hydrogen peroxide in the solution.

All these results concur to exclude the involvement of hydrogen peroxide as a co-product of the photo-oxidation of water and intermediate of other photo-oxidation reactions at TiO_2. The high rates of its photodecomposition at open circuit and photo-oxidation under anodic bias render unlikely the build-up of any significant concentration of H_2O_2 at the surface of irradiated titanium dioxide or in the nearby solution.

In addition to the photoelectrochemical (catalysed by TiO_2) mechanism discussed above, also the chemical and photochemical mechanisms of H_2O_2 decomposition are to be considered. In strongly irradiated, alkaline solutions the direct photolysis of hydrogen peroxide (strictly speaking-of HO_2^- ions) should seriously be taken into account for the

---

17 Capacity measurements (represented as Mott-Schottky plots), performed by Ferrer et al.[170] for a thin film TiO_2 electrode immersed in a less alkaline (pH 11.3) Na_2SO_4 solution, have shown that the addition of H_2O_2 caused effectively the shift of the flat-band potential closely similar to that of the onset potential in Fig. 12.

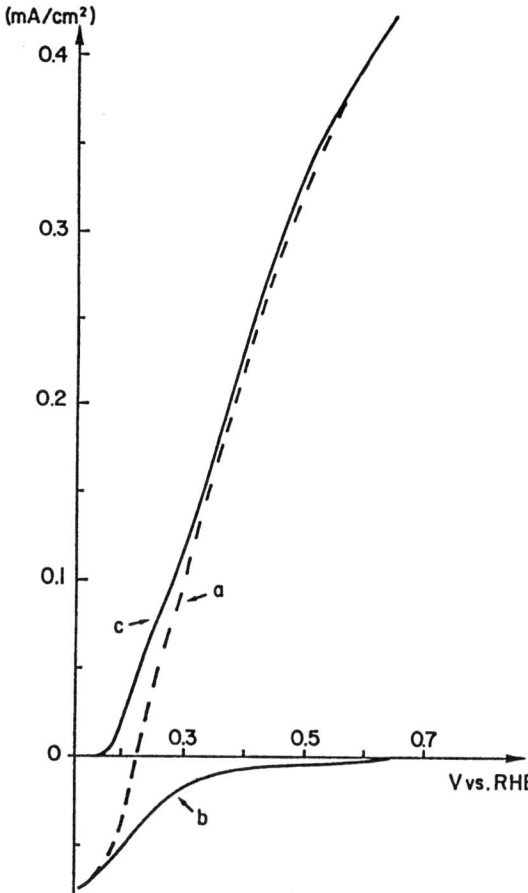

Fig. 13. Polarization curves (recorded at 1 mV · s^{-1}) corresponding to the photoanodic oxidation and the cathodic reduction of hydrogen peroxide at a polycrystalline TiO$_2$-film electrode. *Curve a* is for an illuminated ($\lambda > 335$ nm) and *curve b* for an unilluminated electrode (see text). Measurements performed in a deaerated 0.1 M NaOH/10^{-3} M H$_2$O$_2$ solution

wavelengths shorter than ca. 320 nm. In fact, the extinction coefficient for HO$_2^-$ ions, determined in H$_2$O$_2$/0.1 M NaOH solutions at 3135 Å, amounts to 12 dm^3 mol^{-1} cm^{-1} [172].

4.3.2 *ESR Spectroscopic Studies of the Species Involved in the Photo-Oxidation Reactions at TiO$_2$*

There have been several attempts to identify the intermediates involved in the photo-oxidation reactions and, in particular, in the photo-oxidation of water at TiO$_2$ by means of electron spin resonance (ESR) spectroscopy.

Jaeger and Bard[173], using a spin trapping technique (with α-phenyl N-tert-butyl nitrone, PBN, and α-(4-pyridyl N-oxide) N-tert-butyl nitrone, 4-POBN, as spin traps), have detected hydroxyl, 'OH, and perhydroxyl, HO$_2$, radicals during band-gap irradiation of aqueous suspensions of TiO$_2$ and platinised TiO$_2$ powders. Titanium dioxide employed was a pre-reduced anatase powder and the experiments were performed in deaerated solutions of pH 4 and 7. The ˙OH radicals, observed under these conditions, have been proposed[173] to originate from the photoanodic reaction

$$OH^- + h^+ \rightarrow \ ˙OH \tag{25}$$

The formal standard redox potential of the ˙OH/OH⁻(H₂O) couple being equal to 2.85 V versus RHE[144], i.e. quite close to the valence band edge of TiO_2, reaction (25) might conceivably occur without mediation of the surface states (obviously, the energetics of reaction (25) would be changed if the OH^- ions or ˙OH radicals, or the both, were chemisorbed on the TiO_2 surface). According to Jaeger and Bard, the second radical detected by them, $HO_2^˙$, is to be rather associated with a reduction process involving photogenerated molecular oxygen.

Similar ESR experiments with TiO_2 aqueous dispersions irradiated with an UV light have also been reported by Harbour et al.[160]. These authors used an apparently untreated anatase powder and a different spin trap: 5,5-dimethyl-1-pyrrolinyl-1-oxy (DMPO). The hydroxyl radical has been detected but no evidence for the presence of perhydroxyl radical was found[160]. Importantly, the intensity of the ESR signal, associated with the DMPO˙OH radical adduct, was a function of the oxygen concentration in the aqueous dispersion and became very small after purging with nitrogen.

Howe and Grätzel[174] have extended the ESR investigations to UV-irradiated colloidal TiO_2 solutions (the colloid particles consisted of a mixture of anatase and amorphous titanium dioxide). The spectra were taken for deaerated acidic and alkaline colloidal solutions frozen at 77 or 4.2 K – without spin trap. The only paramagnetic species detected by Howe and Grätzel were the surface and interstitial Ti^{3+} ions. In the deaerated solutions containing an efficient hole scavanger (such as, e.g., methanol) both of these species were stable at room temperature. On the other hand, in the absence of an added hole scavanger only interstitial Ti^{3+} ions could be observed for the samples irradiated at low temperature, i.e., at 4.2 or 77 K. No species associated with the hole trapping at TiO_2, such as ˙OH radicals, were detected in the latter case, even at 4.2 K.

Anpo et al.[175] have performed ESR and photoluminescence experiments with small (having a diameter of 50 to 500 Å) titanium dioxide particles irradiated at 77 K. These particles had anatase structure and, judging by the method employed for their preparation, contained both chloride and sulphate impurities. A prolonged UV irradiation of the TiO_2 powder samples, containing a few monolayers of preadsorbed water, gave rise to a quite intense ESR signal, assigned to adsorbed ˙OH radicals, and to a less intense signal due to Ti^{3+} ions[175]. The latter samples exhibited also photoluminescence, with a distinct maximum at ca. 500 nm, which has been interpreted by Anpo et al. in terms of radiative recombination of the Ti^{3+}–˙OH pairs, photogenerated at the surface of TiO_2 subjected to the band-gap irradiation. The intensity of the observed photoluminescence increased with increasing the amount of water preadsorbed on the surface of TiO_2 but rapidly decreased in the presence of already small quantities of oxygen.

The assignment by Anpo et al.[175] of the ESR signal characterized by $g_1 = 2.014$ and $g_2 = 2.003$ to the ˙OH radicals has been questioned in a recent paper by Howe and Grätzel[176]. Following the earlier ESR study of colloidal TiO_2 dispersions[174], these authors have investigated in situ the paramagnetic species formed on band-gap irradiation of a large-surface-area hydrated anatase powder. Their ESR measurements, performed at 4.2 and 77 K, showed the presence of a signal closely similar to that attributed by Anpo et al.[175] to photogenerated ˙OH radicals. However, according to Howe and Grätzel[176] the observed anisotropic ^1H hyperfine couplings (of less than ca. 5 gauss) are much too small to be consistent with the ˙OH species. The latter authors[176] postulate the observed ESR signal to be associated with ˙O⁻ radical anions resulting from trapping of positive holes at lattice oxide ions. The corresponding trap sites are suggested to be located in the immediate sub-surface layer of anatase, leading to the

$$Ti^{4+}O^-Ti^{4+}OH^-$$

structure. The important conclusion of Howe and Grätzel[176] is that hydroxyl radicals, detected at TiO_2 in spin trapping experiments[173, 160], are not the primary product of hole trapping but arise as transient intermediates of the photo-oxidation process.

Neither the ESR investigations discussed above nor similar earlier works[177] lead to a clear picture of the radical intermediates involved in the photo-oxidation reactions at titanium dioxide. In fact, the nature of the observed radicals appears to depend on the origin and pretreatment of the sample, eventually on the conditions and extent of its reduction, on the degree of hydroxylation of the surface, on the presence of oxygen and other electron acceptor or donor species. It is to be recalled in this connection that, as discussed in Sect. 2.1, a high temperature degassing of the TiO_2 sample usually leads to irreversible changes in its surface structure and, especially, to its irreversible dehydroxylation.

Various studies concerned with the photoadsorption of oxygen at more or less dehydroxylated TiO_2 powders allowed to identify a series of additional paramagnetic species, including O^- [178, 179], O_2^- [180–183], O_3^- [179–181] and O_3^{3-} [180, 181]. The O_2^- species have also been shown to be formed in the absence of illumination – as a result of oxygen adsorption on the surface of reduced TiO_2[184, 185].

4.3.3 The Origin of the Surface Peroxo Titanium Species

The evidence, provided by the voltammetric measurements[137], that the surface peroxo titanium species are formed through the photoanodic pathway, involving positive holes photogenerated in the valence band of TiO_2, designates surface hydroxyl groups as the most plausible reacting species. As specified in Sect. 2.4, the fully hydroxylated anatase and rutile contain ca. 5 surface OH groups per nm^2, including half of the acidic (doubly co-ordinated) and half of the essentially basic (singly co-ordinated) hydroxyls[34] (cf. Fig. 1). About 90 per cent of all the OH groups appear to be subjected to neutralization adsorption of alkaline (or alkaline earth) hydroxides. For example, in the case of a NaOH solution:

$$Ti_s-OH + NaOH \rightleftarrows Ti_s-O^-Na^+ + H_2O \tag{26}$$

where Na^+ denotes an adsorbed sodium cation located in the outer Helmholtz plane (see also Eqs. (3) and (7)).

From the adsorption isotherm established by Boehm[34] for an anatase powder (P-25), the number of charged (Ti_s-O^-) groups, in equilibrium with a 0.1 M NaOH solution, may be evaluated as equal to ca. 3.8 groups/nm^2 [137]. The capture of the photogenerated positive holes by these groups would lead to the formation of $Ti_s{-}O^\bullet$ radicals (cf. Eq. (24)).

It appears necessary, in this connection, to point out that the formula which symbolizes in Eq. (24) a dissociated surface OH group, $Ti_s{-}O^-$, is rather unlikely to reflect the true nature of its bonding to the surface Ti^{4+} ion. To account for the fact that the actual bonding is probably not totally ionic[36], the hydroxyl groups, resulting from the dissociative adsorption of water at TiO_2, should more properly be represented using fractional charges (cf. Eqs. (1) and (2)). This implies that the $Ti_s{-}O^\bullet$ radical should, in fact, be regarded as $^\bullet O^-$ species strongly interacting with the TiO_2 surface.

The photogenerated $Ti_s{-}O^\bullet$ radicals may, in principle, act as (i) surface recombination centres for the conduction band electrons, (ii) electron acceptors towards reduced species present in the solution, or, for example, (iii) undergo conproportionation

$$2\,(Ti_s{-}O^\bullet) \rightarrow Ti_s{-}O{-}O{-}Ti_s \tag{27}$$

to form surface-bonded μ-peroxo species[135, 137, 169].

Significantly, from the results of picosecond and nanosecond flash photolysis experiments, performed with aqueous dispersions of colloidal TiO_2 particles[186], it appears that the trapped holes (i.e., the $Ti_s{-}O^\bullet$ and/or $Ti_s{-}O{-}O{-}Ti_s$ species) react rather slowly with trapped electrons. Because of a large difference between the rates of electron and hole trapping, requiring an average time of only 30 ps for the former and of 250 ns for the latter[186], the photoinduced charge recombination seems to involve principally the trapped electrons and the free holes. This explains why the surface coverage of the irradiated TiO_2 electrodes or particles by the peroxo species may attain the observed large steady-state values.

First quantitative assessment, based on the results of spectrophotometric titrations with $KMnO_4$[133], indicated a concentration of ca. 1.4 peroxo groups per nm^2 of the true

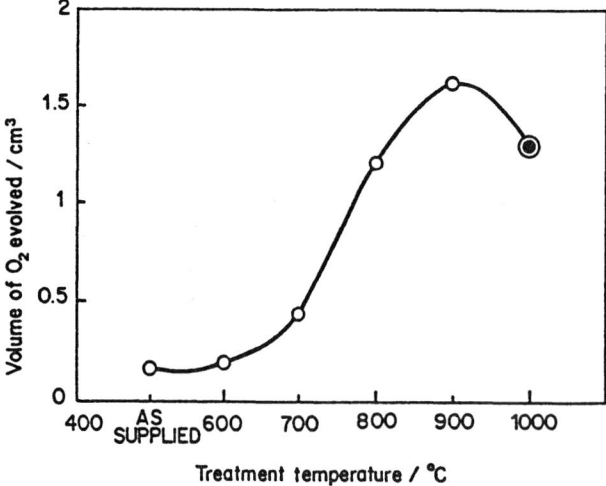

Fig. 14. Volume of oxygen evolved during 5 h of photolysis over TiO_2 powder pretreated at various temperatures (under argon). 100 mg of TiO_2 rutile dispersed in 25 cm^3 of solution containing 10^{-3} mol of $AgNO_3$ (from Ref. 188)

surface area of the Rh-loaded anatase P-25 powder which had been used as a photocatalyst in a 40-h water photolysis experiment. The latter value corresponded to the part of peroxides detected in the Rh(0.3%)/TiO$_2$ catalyst, after its separation from the solution (1 M aqueous NaOH), and constituted probably an underestimate as the peroxides tend to decompose in the course of titration[133]. In fact, recent spectrophotometric analyses using o-tolidine (instead of permanganate) as a redox indicator have shown[187] that during photolysis of water over a TiO$_2$ photocatalyst the hydrogen evolution was accompanied by the formation of a stoichiometric amount (in a 1:1 molar ratio) of surface-bonded peroxo species. The latter experiments were performed with Pt-loaded anatase P-25 particles dispersed in a solution having an initial pH of 10. Under these conditions, the total amount of photogenerated peroxide was observed to be associated with the TiO$_2$ photocatalyst. The saturation coverage, determined after the dispersion had been illuminated for 48 h with a solar simulator, amounted to ca. 4.6 peroxo groups per nm^2.

In the case of a TiO$_2$ photoelectrode the surface coverage of the peroxo species photogenerated under anodic bias may be estimated from the corresponding cathodic voltammograms, like those shown in Fig. 5. Thus, the amount of cathodic charge associated with curve d (Fig. 5), which apparently does not yet represent a saturation coverage, exceeds 10^{17} electrons per cm^2 of geometrical surface area of the electrode. Dividing the latter value by the approximate surface roughness factor ranging from 120[18] to 200[137] and by the number of electrons involved in the reduction of a peroxo group, 2 (cf. Eq. (10)), one obtains 2.5, respectively, 4.2 peroxo groups per real nm^2.

It should also be mentioned that estimates of the concentration of the surface peroxo species, based on the amount of oxygen photoadsorbed at TiO$_2$ powders dispersed in aqueous solution, may lead in some cases to extremely large formal coverages, as the value of 22 peroxide molecules per nm^2 reported by Harbour et al.[160]. This would suggest the formation of three-dimensional peroxide network if it were effectively confirmed that in the latter case the total amount of photoadsorbed oxygen was present in the form of peroxides.

The essential point arising from the above discussion is that the saturation coverage of titanium dioxide with the surface-bonded peroxo species, photogenerated in alkaline or neutral solutions, ranges most likely from 4 to 5 peroxo groups per nm^2. The latter value is close to the total number of active OH groups present initially on each nm^2 of fully hydroxylated (and unilluminated) anatase surface.

The involvement of the surface OH groups in the formation of peroxo titanium species has recently been confirmed by experiments performed by Oosawa and Grätzel[188] using a TiO$_2$ rutile photocatalyst pretreated at different temperatures ranging from 873 to 1273 K (such pretreatments are known to cause a more or less irreversible dehydroxylation of titanium dioxide; cf. Sect. 2.4). As shown in Fig. 14, the amount of oxygen formed during photolysis of an aqueous silver nitrate solution

$$4\,Ag^+ + 2\,H_2O \xrightarrow{\ h\nu\ } 4\,Ag(TiO_2) + 4\,H^+_{aq} + O_2 \qquad (28)$$

---

18 The value of 120 has been calculated from the amount of charge associated with the voltammetric reduction of hydrogen peroxide preadsorbed at the TiO$_2$ electrode (cf. Fig. 11), using the H$_2$O$_2$ adsorption isotherm determined by Boonstra and Mutsaers[145] for an anatase powder sample.

was strongly affected by the pretreatment temperature of the rutile powder, increasing about ten times for TiO_2 heated at 1173 K with respect to its untreated equivalent. Also the molar ratio of silver to oxygen, produced through the above photoreaction, was markedly influenced by the initial degree of hydroxylation of the TiO_2 powder. It varied between the stoichiometric value of 4, for practically dehydroxylated TiO_2, and 6.6 for untreated photocatalyst. The formation of peroxo species on the surface of hydroxylated TiO_2 particles allows one to explain not only the deficit of oxygen but also the observed low rate of silver deposition at the untreated photocatalyst. The latter result may be interpreted as due to the increased hole-electron recombination through the formation and reduction of the surface peroxo species. An additional conclusion, arising from the above work, is that the $Ti_s–O^{•}$ radicals and/or peroxo-titanium species are probably not the unique nor necessary intermediates of the photo-oxidation of water.

Considering the possible structure and mechanism of formation of the peroxo species consistent with the observed coverage of the TiO_2 surface, i.e. 4–5 molecules per nm^2, one notes immediately that the pairing of the $Ti_s–O^{•}$ radicals, according to Eq. (27), might only account for a coverage slightly higher than one $Ti_s–O–O–Ti_s$ group per nm^2. In fact, the reaction (27) should, in principle, involve singly co-ordinated hydroxyls (cf. Fig. 1), the number of which does not exceed an average of 2.5 groups per nm^2. In addition, these basic OH groups are expected to become entirely dissociated only in strongly alkaline solutions. Therefore, it seems more likely, especially for titanium dioxide irradiated in neutral or slightly alkaline solutions, that the hole capture occurs at doubly co-ordinated (acidic) OH groups

$$
\begin{array}{ccc}
\text{OH} & \text{OH} & \\
| & | & \\
Ti_s–O^{-}–Ti_s & + \ h^{+} & \longrightarrow
\end{array}
\begin{array}{cc}
\text{OH} & \text{OH} \\
| & | \\
Ti_s–O^{•}–Ti_s
\end{array}
\tag{29}
$$

This initial step may be followed by a second charge-transfer reaction

$$
\begin{array}{cc}
\text{OH} & \text{OH} \\
| & | \\
Ti_s–O^{•}–Ti_s & + \ h^{+}
\end{array}
\longrightarrow
\begin{array}{cc}
\text{O} & \text{OH} \\
/ \ \backslash & | \\
Ti_s–O–Ti_s & + \ H^{+}_{aq}
\end{array}
\tag{30}
$$

leading to the formation of species with one peroxo group per titanium ion.

The maximum concentration of the latter species should correspond to about 2.5 peroxo groups per nm^2, i.e. half of what is assumed to be the total coverage of active OH groups.

Even if it seems well established that the surface hydroxyls play an essential role in the photogeneration of the peroxo-titanium species, this does not preclude that water molecules and/or OH^{-} ions from the solution may simultaneously be also involved in such a process. This can, for example, occur through reaction (29) followed by

$$
\begin{array}{cc}
\text{OH} & \text{OH} \\
| & | \\
Ti_s–O^{•}–Ti_s & + \ H_2O + h^{+}
\end{array}
\longrightarrow
\begin{array}{ccc}
\text{OH} & \text{OH} & \text{OH} \\
| & | & | \\
Ti_s–O–\!\!—Ti_s & + \ H^{+}_{aq}
\end{array}
\tag{31}
$$

In strongly alkaline solutions also the singly co-ordinated (basic) OH groups may presumably react in a similar way, i.e.,

$$
\begin{array}{ccc}
\mathrm{O^-} & & \mathrm{O^\bullet} \\
| & & | \\
\mathrm{Ti_s} & + \ \mathrm{h^+} \ \longrightarrow & \mathrm{Ti_s}
\end{array}
\tag{32}
$$

and

$$
\begin{array}{cccc}
& & & \mathrm{OH} \\
& & & | \\
\mathrm{O^\bullet} & & & \mathrm{O} \\
| & & & | \\
\mathrm{Ti_s} & + \ \mathrm{OH^-} \ + \ \mathrm{h^+} \ \longrightarrow & & \mathrm{Ti_s}
\end{array}
\tag{33}
$$

The formation of such linear peroxo species

$$
\begin{array}{ccc}
\mathrm{OH} & & \mathrm{OH} \\
| & & | \\
\mathrm{O} & \mathrm{OH} & \mathrm{O} \\
| & | & | \\
\mathrm{Ti_s} - & \!\!\!\mathrm{O} - \!\!\! & \mathrm{Ti_s}
\end{array}
\tag{34}
$$

would eventually account, especially in alkaline solutions, for the coverages close to those observed experimentally.

It is to be noted that both reactions (30) and (31) lead to the release of protons into the solution. This is consistent with a perceptible decrease of pH occurring during the photo-uptake of oxygen[133] or, e.g., the photodeposition of platinum[187] in quasi-neutral or slightly alkaline aqueous dispersions of TiO_2.

Unlike that of the surface hydroxyl groups, the role which the molecular water, adsorbed at titanium dioxide, may possibly play in the photogeneration of the peroxo-titanium species remains rather obscure. Still, some indications are provided by temperature-programmed desorption (TPD) profiles of hydroxylated TiO_2 samples subjected to a prolonged band-gap irradiation. While an unilluminated TiO_2 exhibits typically three water desorption peaks (cf. Sect. 2.1), the two higher-temperature peaks, assigned to surface OH groups, are no more observed for the sample irradiated with the UV light, presumably due to the conversion of surface hydroxyls into the peroxo species[189]. On the other hand, the lowest-temperature desorption maximum, associated with the adsorbed molecular water, remains practically unaffected by the sustained illumination[189]. This would suggest only marginal involvement of the adsorbed water in the formation of the surface-bonded peroxo species.

4.4 Mechanism of the Oxygen Photogeneration at TiO_2

The presence of the peroxo compounds, bound to the surface of titanium dioxide, appears as the permanent feature of the photoanodic reactions involving oxidation of water and/or hydroxyl ions. In the case of aqueous TiO_2 suspensions irradiated with UV light, the surface-bonded peroxo species may constitute practically the only product of the photo-oxidation reaction. This occurs, in particular, in closed systems, operating at room temperature and under atmospheric pressure, when the cathodic reaction consists in the reduction of water into hydrogen. Under such conditions, the amount of photogenerated hydrogen is in fact controlled by the maximum surface coverage of the TiO_2 photocatalyst with the peroxo species. The limiting role of the surface coverage is illus-

trated by the observations of Gu et al.[135] regarding the effect of Ba^{2+} ions upon efficiency of the photochemical splitting of water in alkaline Pt/TiO_2 suspensions. Addition
of $0.05 \, mol \cdot dm^{-3}$ of barium hydroxide to 0.1 M NaOH solution caused a substantial
(more than twofold) enhancement of the total amount of photogenerated hydrogen
(Fig. 15). As shown in Fig. 15, the influence of Ba^{2+} cations upon the initial H_2 formation rate was even more marked, the latter value being ca. six times larger for the
photocatalyst suspension containing $Ba(OH)_2$. The observed effects have been
ascribed[135] to the ability of Ba^{2+} ions to form insoluble barium peroxide, decreasing the
concentration of the peroxo species at the surface of the photocatalyst. In addition, one
has to take into account that the divalent barium ions, adsorbed at acidic OH groups on
the TiO_2 surface, may be directly incorporated into the photogenerated peroxo-titanium
species. Such modified surface peroxides are very likely to be less easily reducible, thus
decreasing the extent of the surface electron-hole recombination.

Unlike in the cases when hydrogen is photogenerated over titanium dioxide photocatalyst, the photoreactions involving reduction of the species having redox potentials
more positive than that of the hydrogen electrode, as for example

$$Ag^+ + e^-(TiO_2) \rightarrow Ag(TiO_2) \tag{35}$$

are, in general, characterized by the simultaneous evolution of molecular oxygen. However, even under these conditions, the surface peroxo species constitute a co-product of
the photoanodic reaction as indicated by the molar ratios of photoproduced oxygen to
silver – lower than the stoichiometric value of $1:4$[188, 190].

These observations are consistent with an indirect mechanism of oxygen evolution,
involving the photogeneration and the subsequent photo-oxidation of the surface peroxo-
titanates. Apparently, the photo-oxidation of the peroxo species constitutes the slow step
of the overall reaction and proceeds at a significant rate only in the presence of an anodic

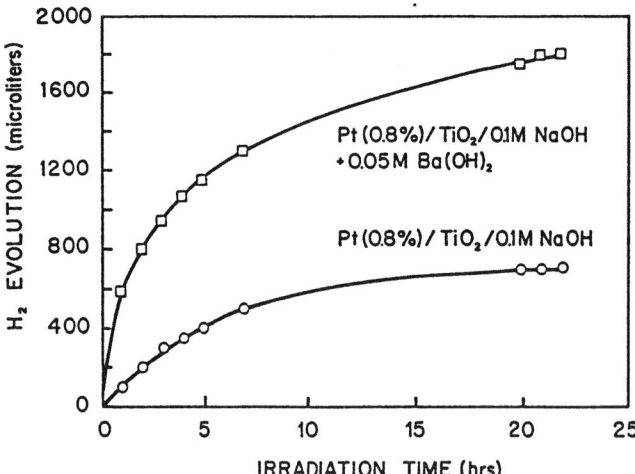

Fig. 15. Photogeneration of hydrogen over aqueous dispersions of $Pt(0.8\%)/TiO_2$. 50 mg of catalyst
in 25 cm^3 of 0.1 M NaOH were irradiated with the $\lambda > 320$ nm output of a 150 W xenon lamp (from
Ref. 135)

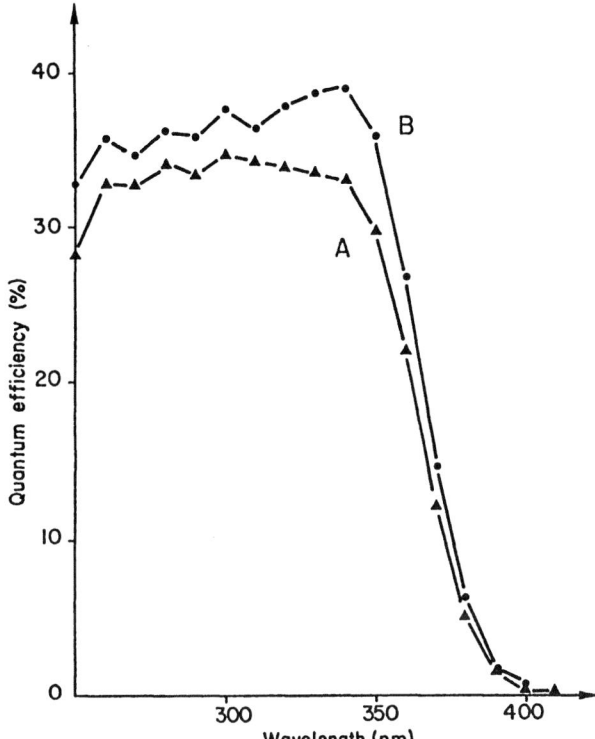

Fig. 16. Spectral photoresponses of a polycrystalline TiO_2 electrode polarized at 0.8 V (vs. RHE) in 0.1 M aqueous NaOH. Curve A is for the electrode prepolarized at 0.8 V under the $\lambda = 340$ nm illumination

overvoltage. In the case of the TiO_2 photocatalyst dispersed in aqueous solution, the overvoltage will depend upon the actual value of the photopotential and, therefore, upon the redox potential of the species undergoing reduction in the course of the photoreaction.

The increase of the photoanodic overvoltage, associated with the coverage of the TiO_2 photoelectrode by the peroxo species, may be directly evaluated from Fig. 16[191]. The prepolarization of the TiO_2 electrode, leading to the steady-state coverage by the surface-bonded peroxides, is shown to cause a perceptible decrease of the quantum efficiency of the photocurrent (curve A). Curve B in Fig. 16 has been obtained under identical conditions and for the same TiO_2 electrode made practically free from the peroxo species. As at this relatively positive potential (about 0.8 V more positive than the flat-band potential of anatase) the surface electron-hole recombination becomes negligible, the observed diminution of the photocurrent (attaining, in Fig. 16A, 15 per cent) is to be ascribed as a whole to the increase of the photoanodic overvoltage.

A plausible way by which the photo-oxidation of the surface peroxo-titanium species may take place is the formation of the corresponding surface superoxo-titanium compound and its subsequent decomposition. It is to be noted, in this connection, that the superoxo complex of titanium (IV), TiO_2^{3+}, has been postulated to be the intermediate in the oxidation of peroxo-titanium (IV), TiO_2^{2+}, by cerium (IV) in perchloric acid solution[192].

$$TiO_2^{2+} + Ce(IV) \rightarrow TiO_2^{3+} + Ce(III) \tag{36}$$

According to the reaction scheme proposed by Thompson[192], the final evolution of molecular oxygen would occur through the decomposition of the superoxo-titanium complex, involving a relatively slow internal redox reaction

$$TiO_2^{3+} \rightarrow O_2 + Ti^{3+} \tag{37}$$

The photo-oxidation of the surface-bonded peroxo-titanates may formally be represented as

$$
\begin{array}{ccccc}
OH & & OH & & O^{\bullet} & & OH \\
| & & | & & | & & | \\
O & OH & O & & O & OH & O \\
| & | & | & & | & | & | \\
Ti_s{-}O{-}Ti_s & + & h^+ & \longrightarrow & Ti_s{-}O{-}Ti_s & + & H_{aq}^+
\end{array}
\tag{38}
$$

followed by the oxygen evolution according to

$$
\begin{array}{ccccc}
O^{\bullet} & & OH & & & & OH \\
| & & | & & & & | \\
O & OH & O & & OH & OH & O \\
| & | & | & & | & | & | \\
Ti_s{-}O{-}Ti_s & + OH^- + h^+ & \longrightarrow & Ti_s{-}O{-}Ti_s & + O_2
\end{array}
\tag{39}
$$

Reaction (39) may be seen either as a direct photo-oxidation of the surface superoxo-titanium species or as their decomposition followed by a rapid reoxidation of the resulting Ti^{3+} site into Ti^{4+}.

The above proposed mechanism of photo-oxidation of water at titanium dioxide, including the sequence of reactions (32), (33), (38) and (39), presents some analogy with the mechanism of anodic oxygen evolution at a palladium electrode, in acid solutions, involving electrochemical oxidation of the metal into unstable palladium dioxide, PdO_2, and its subsequent decomposition[193].

The question arises whether the indirect mechanism, involving the formation of surface peroxo-titanium and superoxo-titanium species, constitutes the only possible way by which molecular oxygen may be photogenerated at titanium dioxide. The results of Oosawa and Grätzel[188] (discussed in Sect. 4.3.3), concerning the stoichiometry of the photoreactions occurring at partially dehydroxylated rutile powders, dispersed in aqueous solution of $AgNO_3$, suggest that the photoproduction of oxygen does not necessarily require a significant coverage of TiO_2 by the peroxo species. However, one has to take into account that the oxygen evolution may, eventually, be catalysed in this case by the photodeposited silver particles. The same remark can be made with regard to the TiO_2 photocatalyst loaded with both ruthenium dioxide and platinum particles. The presence of RuO_2, known as an efficient electrocatalyst of oxygen evolution, apparently favours simultaneous photogeneration of oxygen and hydrogen. Still, the reported O_2/H_2 molar ratio close to $1:3$[18] would indicate that even in this case a non negligible portion of the photoinduced positive holes are used to form the surface-bonded peroxo species.

5 Photo-Oxidation Reactions at the TiO$_2$ Photoanode

As a rule, the photo-oxidation of the redox species, present in an aqueous solution, occurs at a TiO$_2$ photoanode simultaneously with that of water and/or hydroxyl ions leading to the oxygen evolution. Two other processes, competing equally for the photo-generated positive holes reaching the surface of the semiconductor, are the surface hole-electron recombination and, in some cases, the photocorrosion. Titanium dioxide being a particularly stable semiconductor, the latter process may be practically neglected except for strongly acidic solutions[194, 195]. The rate of the recombination will depend on the simultaneous availability of the minority and majority charge carriers at the semiconductor surface. Consequently, the recombination will be particularly effective on the surface of a semiconductor irradiated with light of strong intensity. The rate of the hole-electron recombination should be the highest at and around the flat-band potential of the semiconductor and decrease with increasing the external bias (and, i.e., the band bending). Consequently, the amount of the saturation photocurrent is no more controlled by the kinetics of the various surface processes but exclusively by the effective light intensity and the solid-state properties of the outermost (in general thinner than 1 μm) layer of the semiconductor[19]. Therefore, it is the analysis of the anodic photocurrents in the range of potentials close to the flat-band potential[20] which may provide useful information about the kinetics of the competitive photoreactions at TiO$_2$.

---

19 When a wide-band-gap n-type semiconductor, such as TiO$_2$, irradiated with monochromatic light, is subjected to a large anodic bias, the resulting photocurrent, flowing through the interface with the solution, may, in general, be described by the relation[196, 197].

$$i_{ph} = e\Phi_0\{1 - [\exp(-\alpha W)](1 + \alpha L_p)^{-1}\} \tag{40}$$

where e is the elementary charge, Φ_0 is the incident photon flux, α is the optical absorption coefficient, W is the width of the depletion layer and L$_p$ is the diffusion length of the minority charge carriers (holes). Both W and L$_p$ depend in the similar way upon the donor concentration N$_d$, i.e.

$$W = (2 \, \varepsilon_r \varepsilon_0 / eN_d)^{\frac{1}{2}} (E - V_{fb})^{\frac{1}{2}} \tag{41}$$

where ε_r is the dielectric constant of the depletion layer, ε_0 is the permittivity of vacuum, $(E - V_{fb})$ is the band bending. The hole diffusion length is given by

$$L_p = [\mu_p/(\mu_p + \mu_e)^{\frac{1}{2}}](\varepsilon_r \varepsilon_0 \, kT/4 \, \pi \, e^2 N_d)^{\frac{1}{2}} \tag{42}$$

where μ_p and μ_e are, respectively, the hole and electron mobilities and k is the Boltzmann constant.
20 The photocurrent determined experimentally in this range of potentials should be thoroughly corrected for the possible contribution arising from the reduction current due to the majority carriers.

5.1 Competition Between the Photo-Oxidation of Various Reducing Species and the Oxygen Evolution

An appropriate example is given by the photocurrent-voltage curves for the photo-oxidation at the titanium dioxide photoanode of hydrogen peroxide added to a NaOH solution (Fig. 12). These results, discussed in some detail in Sect. 4.3.1, are a good illustration of significant differences occurring in the rates of hole transfer to OH⁻ ions (or water) and to other compounds in the solution. Very similar observations have also been made as regards the photo-oxidation of methanol and sodium formate from 0.1 M aqueous NaOH[191]. As shown in Fig. 17A and B, the presence of these species in the solution affects especially the initial part of the photocurrent-voltage curves, resulting in an important enhancement of the photocurrent and the shift of its onset to more negative potentials. This kind of behaviour is characteristic of the species undergoing rapid charge (hole) transfer at a semiconductor photoanode[168], and thus able to compete efficiently with the surface recombination.

The ratio of the (partial) photocurrents, due to the photo-oxidation of two species (M and W) competing for holes at the surface of a photoanode, may be formally expressed as[118]

$$i_{pM}/i_{pW} = \delta_M[M]/\delta_W[W] \tag{43}$$

where the photocurrent associated with each species is considered to be proportional to their capture cross section, δ, and to their true surface density.

Even if all the above mentioned species, i.e., H_2O_2[145], CH_3OH[34] and $HCOONa$[198] are known to interact rather strongly with titanium dioxide, their effective surface densities in equilibrium with relatively dilute (0.01 M) solutions can hardly be expected to exceed the surface concentration of OH groups and water molecules adsorbed at TiO_2. For instance, the coverage of hydrogen peroxide, adsorbed from its 0.01 M solution, reaches an average of 1.4 molecules per nm^2 of true surface area of anatase as compared with the surface concentration of ca. 5 active OH groups per nm^2. Thus, the marked increase of the anodic photocurrents, observed for the solutions containing hydrogen peroxide, methanol or sodium formate, is indicative of the large hole capture cross sections of these compounds – larger than for OH⁻ ions.

In this connection it is useful to specify that the capture cross section, which describes the reactivity of the species in the solution towards holes photoproduced in the semiconductor, may have not the same physical meaning for various reactions. In fact, because of the highly exothermic nature of the hole transfer from the valence band of TiO_2 (placed at ca. 3 V versus RHE) to redox couples stable in aqueous solution, such processes are generally considered to be mediated by the surface states located into the band gap. In particular, hydroxyl groups on the titania surface have frequently been mentioned as possible hole traps[199, 118] (cf. reaction (24)).

Now, let us assume that the transfer of holes, photogenerated in the valence band of TiO_2, to the species undergoing photo-oxidation proceeds, in most cases, in two steps, for example, the hole capture at the Ti_s–O^- site[21] followed by the electron exchange

---

21 The example quoted herein is relevant to the photo-oxidation reactions occurring in alkaline solutions (cf. Sect. 4.3.3). In acidic solutions, similar role is probably played by the undissociated Ti_s–OH species.

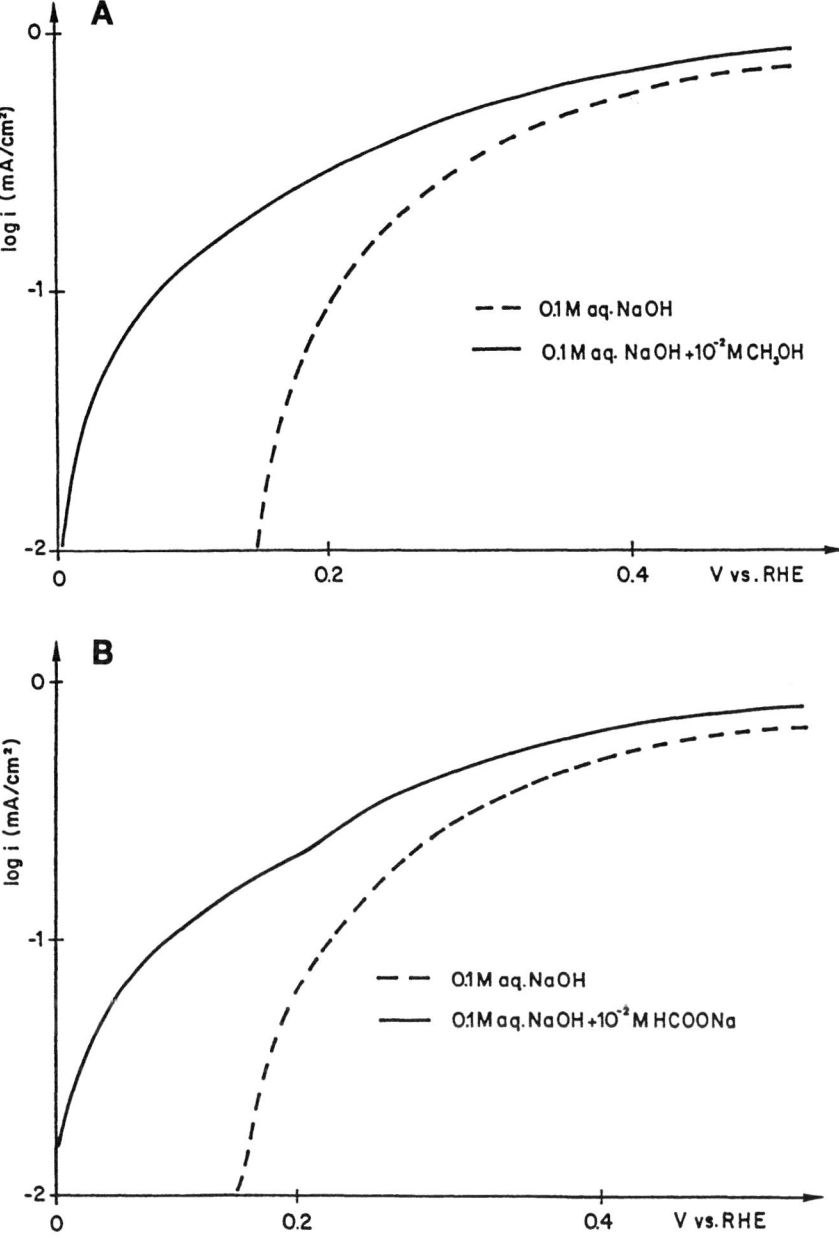

Fig. 17A, B. Effect of the addition of methanol (part **A**) and sodium formate (part **B**) upon photo-current-voltage curves for a TiO$_2$ film electrode in 0.1 M NaOH. Electrode illuminated with the $\lambda > 335$ nm output of a 150 W Xe lamp

between $Ti_s-O^{\cdot}$ and the reactant present in the solution. This would imply that the differences observed between hole capture cross sections of various reducing species are associated with the second step of the reaction.

In particular, the photoanodic oxidation of iodide, bromide and chloride ions at titanium dioxide is very likely to follow such a two-step mechanism. The current efficiency, with which halide ions compete for photogenerated holes, has been observed to increase in the order of decreasing redox potentials, i.e., $Cl^- < Br^- < I^-$ [200-205]. Such a sequence of preferential photo-oxidation is opposite of that expected on the basis of the fluctuating energy level model for direct hole transfer from the valence band of TiO_2 to the ions in the solution[118]. In fact, according to this model, the rate of hole capture should decrease exponentially with increasing $(V_{vs} - E_{red})^2$, where V_{vs} is the potential of the valence band edge at the surface of the semiconductor and E_{red} is the most probable energy level for a reducing species (expressed with respect to the same reference electrode as V_{vs}).

The experimental results regarding the photo-oxidation of halides at TiO_2 strongly suggest that these reactions proceed through an indirect, two-step pathway, involving an initial hole capture by surface states located most probably close to the mid band gap, i.e., at ca. 1.5 V versus NHE at pH 0. Such a position of the surface states would permit a particularly good overlap of the I^- donor levels in the solution (the standard redox potential of the I_2/I^- couple is 0.536 V and we assume here a reorganization energy, λ, of about 0.8-1 eV). The overlap would be less favourable for the Br_2/Br^- and Cl_2/Cl^- couples having standard redox potentials of 1.066 V, respectively, 1.358 V (the ionic radii of I^-, Br^- and Cl^- being similar, there is no reason to expect drastic changes in the corresponding λ values).

Extended comparative measurements of the current efficiencies for the photo-oxidation of a large variety of compounds, in competition with the oxygen evolution, have essentially confirmed the same general tendency for the decrease of the oxidation rate with increasing the standard potential of the redox couple[200, 201, 203]. The highest current efficiency for the competitive photo-oxidation at TiO_2 in acid solutions has been reported for hydroquinone[201, 203] which (i.e., the quinone/hydroquinone couple) has the standard redox potential slightly more positive than that of the I_2/I^- couple.

Although the specific adsorption of anions has been invoked in connection with their competitive photo-oxidation at a TiO_2 photoanode[200] it appears, on the basis of the available data, unlikely to affect the observed order of reactivities. In particular, none of the three halide ions (I^-, Br^-, Cl^-) undergo specific adsorption at titanium dioxide (cf. Sect. 3.2). Actually, the adsorbability of the iodide ion, subjected to the preferential photo-oxidation at TiO_2, is even lower than that of the chloride. This, of course, does not preclude the equivalent (non-specific) adsorption of the above anions, occurring at positively charged TiO_2 surface (i.e., in the solutions of pH lower than the PZC), which may contribute to the increase of current efficiency for the photo-oxidation of iodides, bromides and chlorides with decreasing the pH[201, 203].

In contrast with the three halide ions, the specific adsorption is certainly involved in the photoanodic oxidation of ferrocyanide, $Fe(CN)_6^{4-}$, which has been observed to be strongly attached to the TiO_2 surface[206]. However, this case does not permit any definitive conclusion with regard to the mechanism of the $Fe(CN)_6^{4-}$ photo-oxidation, as the actual high current efficiency for this reaction is, in fact, expected from the position of the standard potential of the $Fe(CN)_6^{3-}/Fe(CN)_6^{4-}$ couple (0.36 V).

5.2 Possible Pathways for the Photo-Oxidation Reactions at TiO$_2$

At first sight, the above mentioned results regarding the competitive photo-oxidation of a series of (mainly inorganic) species may be explained in terms of a general model involving the initial hole capture by the surface states, located close to the mid band gap of TiO$_2$, and the subsequent electron exchange with the reducing species present in the inner Helmholtz plane. Indeed, such a model accounts reasonably well for the observed sequence of preferential photo-oxidation at TiO$_2$ if one assumes for the corresponding reducing species the values of the reorganization energies of the order of 1 eV.

It is, however, to be pointed out that this simplified picture constitutes at the utmost a rough approximation to the real situation. In fact, a close examination of reorganization energies, derived from different series of experimental data[121, 207, 208], shows large differences between λ values of various ions[22] and even between those concerning the same ionic species. Thus, for instance, the values of λ mentioned in the literature for the ferricyanide/ferrocyanide redox couple vary from 0.4 to about 1.2 eV[207, 208].

Moreover, the fluctuating energy level model[209–213], serving as a basis for the above discussion, actually permits to represent the "outer-sphere" electron transfer reactions (i.e. involving neither breaking nor formation of chemical bonds). This means, therefore, that the model should not be used to interpret the photo-oxidation behaviour of water, I$_2$/I$^-$, Br$_2$/Br$^-$, Cl$_2$/Cl$^-$ and many other redox couples. There are, in fact, very few examples, among the reducing species which had been investigated at the TiO$_2$ photoanode, which lend themselves to an analysis in terms of the fluctuating energy level model (even the case of the Fe(CN)$_6^{3-}$/Fe(CN)$_6^{4-}$ couple is doubtful because of the chemisorption of the ferrocyanide ion at titanium dioxide).

An additional complication is associated with badly defined nature of the surface states expected to be involved in the hole transfer processes. In fact, the question arises whether the position of these surface states changes or not with pH of the solution, i.e., whether it follows or not the pH-induced shift of potentials of the valence and conduction band edges at the surface of TiO$_2$. The lack of sufficient data on the pH dependence of the current efficiencies for the "outer-sphere" photo-oxidation reactions at TiO$_2$[23] does not permit to answer directly this question.

Nakato et al.[214, 215] have attempted to identify the surface states, acting as intermediates of the photo-oxidation of water (and, probably, of other species) at TiO$_2$, by measuring the photoluminescence spectra. They observed a sharp photoluminescence peak at 840 nm (1.47 eV) arising when a TiO$_2$ electrode, irradiated with an incident near-UV light, was polarized anodically close to the flat-band potential. Importantly, an identical transient photoluminescence signal was also emitted from the TiO$_2$ electrode which had first been subjected to an anodic bias under UV illumination and then its potential was swept, in the dark, in the cathodic direction. The appearance of the photoluminescence coincided with that of the transient cathodic current occurring near the onset potential of the (anodic) photocurrent. The transient photoluminescence peak at 840 nm was

---

22 In fact, these differences exceed those expected from the variation of the ionic radii.
23 The only available data concern the pH dependence of current efficiencies for the photoanodic oxidation of bromide and other halide ions. These results indicate a decrease of the current efficiency with increasing the solution pH[201, 203].

observed in the latter case even when the electrode had been kept in the dark, for several minutes, before starting the cathodic sweep.

The interpretation of the above photoluminescence spectra, proposed by Nakato et al., involved the capture of the photogenerated holes by the OH^- ions, leading to the formation of a kind of 'OH species, followed by their radiative recombination with the conduction band electrons. To account for the apparently long lifetime of these oxidized intermediates, the authors have suggested them to be located not at the surface but in the bulk of the TiO_2 electrode, within a region of several nanometers from the surface[215]. While the existence of such interstitial 'OH radicals cannot a priori be excluded, the concept of "sub-surface states" acting as intermediates in the process of electron exchange with the reducing species, present in the Helmholtz layer, is all but evident.

An alternative and much simpler assignment for the photoluminescence signal at 840 nm may be deduced from the comparison of the observations of Nakato et al.[215] with those regarding the electrochemical behaviour of the surface-bonded peroxo species[137] described in Sect. 4. The appearance of the transient photoluminescence peak during the cathodic sweep, in the potential region corresponding to the reduction of the peroxo-titanate species, suggests, in fact, that it is the intermediate formed in the course of electrochemical reduction which gives rise to the radiative recombination with the conduction band electron. The first step in the cathodic reduction of surface-bonded peroxo species may be represented schematically as

$$Ti_s-O-O-Ti_s + e^- \rightarrow Ti_s-O^- + Ti_s-O^{\cdot} \tag{44}$$

The radical intermediate, $Ti_s-O^{\cdot}$, can either undergo further cathodic reduction

$$Ti_s-O^{\cdot} + e^- \rightarrow Ti_s-O^- \tag{45}$$

or recombine radiatively with the electron from the conduction band

$$Ti_s-O^{\cdot} + e^- \rightarrow Ti_s-O^- + h\nu \, (1.47 \text{ eV}) \tag{46}$$

(Unlike reaction (46), reaction (45) is associated with the passage of cathodic current in the external circuit.)

As the $Ti_s-O^{\cdot}$ radical is, at the same time, the precursor to the long-lived peroxo-titanate species (cf. reactions (24) and (27)), reaction (46) accounts also for the photo-luminescence at 840 nm observed by Nakato et al.[214, 215] for the TiO_2 electrode irradiated with near-UV light and subjected to moderate anodic bias.

The occurrence of similar luminescence bands, having the maxima at 840 nm, for the TiO_2 electrodes immersed both in acidic and in alkaline solutions implies that the position of the vacant level, involved in the corresponding electronic transitions, with respect to the conduction band level remains unchanged. Accordingly, since the potential of the conduction band edge at the surface of TiO_2 decreases of approximately 0.059 V per pH unit on going from acidic to alkaline solutions, the potential level corresponding to these surface states, taken as equal roughly to 1.5 V versus the normal hydrogen electrode, NHE, at pH 0, will become of about 1.1 V in neutral solution and 0.7 V at pH 14. The knowledge of the influence of pH upon the photo-oxidation behaviour of the reducing species, undergoing outer-sphere charge transfer at the TiO_2 photoanode, would permit to check whether the Ti_s–$O^\cdot$ radicals (and their acidic equivalents) play there effectively the role of intermediates. Still, the strong quenching of the photoluminescence at 840 nm, caused by the addition to a H_2SO_4 solution of hydroquinone[214, 215], suggests that the photo-oxidation of the latter compound is actually mediated by Ti_s–$O^\cdot$ species[24].

On the other hand, the case of cerium (III) ions requires probably a different interpretation. In fact, the positive standard redox potential of the Ce^{4+}/Ce^{3+} couple (1.61 V), associated with a large value of the reorganization energy ($\lambda = 1.7$ eV^{208})) suggest the possibility of a direct electron exchange with the valence band of TiO_2. Such a pathway would account better for the observed high current efficiency for the photoanodic oxidation of the Ce^{3+} ions (reaching 0.7 in the 0.1 M solution)[200] than that involving Ti_s–$O^\cdot$ type intermediates.

In spite of a large number of investigations devoted to the multi-step photo-oxidation reactions at TiO_2, their mechanisms are far from being completely understood. Importantly, many of the compounds, losing several electrons in the course of anodic oxidation, nevertheless compete efficiently with water and/or OH^- ions for the positive holes photogenerated in TiO_2. Some among these compounds, such as above mentioned hydrogen peroxide, methanol or formates, are known moreover to undergo strong adsorption on titanium dioxide, forming in this way the surface states. Depending on their density and the position in the band gap, these surface states may be able to capt directly the holes from the valence band. A well-documented example of such a "direct" hole transfer mechanism is provided by the photo-oxidation of hydrogen peroxide, observed at colloidal TiO_2[169], and is discussed in detail in Sect. 4.3.1.

Several among the compounds undergoing multi-step photo-oxidation at titanium dioxide give rise to a phenomenon defined generally as "photocurrent doubling". This kind of behaviour has been observed for the first time by Morrison and Freund[218] at the single-crystal ZnO photoanode in the presence of formate ions in the solution. The addition of a significant amount (0.1 M $\cdot$ dm^{-3}) of formate to a slightly alkaline KCl solution resulted in almost doubling of the saturation photocurrent at the ZnO photoelectrode[218]. As the amount of the photocurrent observed at large anodic bias is, in principle, expected to depend only on the effective intensity of the incident light (assuming that the characteristics of the space-charge layer are unaffected by the anodic polarization), in order to explain the multiplication of the current Morrison and Freund[218] proposed a two-step valence band-conduction band pathway. This, now widely accepted, mechanism involves, in the case of an n-type semiconducting photoelectrode, an initial transfer of a hole, photogenerated in the valence band, to, e.g., the formate ion

$$HCOO^- + h_{vb}^+ \rightarrow HCOO^\cdot \tag{47}$$

followed by the electron injection by the formyloxy radical to the conduction band

$$HCOO^\cdot \rightarrow H_{aq}^+ + CO_2 + e_{cb}^- \tag{48}$$

---

24 More properly, the above remark refers to the initial step of this reaction. Studies performed using platinum[216] and SnO_2[217] electrodes indicate that the quinone/hydroquinone redox reaction involves two distinct, consecutive charge transfer steps. Also the hydroquinone oxidation at the TiO_2 photoanode follows presumably a two-step mechanism.

resulting in the "current doubling". This implies that the radical species formed in the first step of the reaction are oxidized much easier (i.e., at less positive potentials) than the original substrate. The proposed mechanism is consistent with two kinds of observations, namely that (i) the one-equivalent reducing species do not cause the multiplication of the photocurrent[219] and (ii) the presence of oxygen or of other oxidizing species lead to a suppression of the "current doubling"[219, 220]. The latter effect was ascribed to the quenching of the radical intermediate by an oxidizing agent[218-220], as for example

$$HCOO^{\cdot} + O_2 \rightarrow H_{aq}^+ + CO_2 + O_2^- \tag{49}$$

The "photocurrent doubling" has been verified for the photo-oxidation at the ZnO photoelectrode of a large variety of organic and also some inorganic species[219-223]. Subsequently, similar observations have been made for other n-type semiconducting photoanodes; cadmium sulfide[224], cadmium selenide[225], strontium titanate[226] and titanium dioxide[204, 227-235]. Rather in contrast with the behaviour of other above mentioned photoelectrodes (i.e., ZnO, CdS and CdSe), in the case of the TiO_2 photoanode the multiplication of the photocurrent appears to be more effective in acidic than in alkaline solutions[204].

Characteristically, the species which undergo rapid photo-oxidation at titanium dioxide quite frequently give also rise to the "photocurrent doubling" (for instance methanol, formate ions etc.). This is likely to be connected with the ability of these species to form chemisorption bonds with the oxide surface. The question arises how to distinguish between the two above mentioned kinds of effects which both produce a more or less important increase of the photoanodic current. As a rule, a significant enhancement of the photocurrent, close to the onset potential, caused by the addition to the solution of a small amount of the reducing species, is quite typical of the high rate of charge transfer (cf. Sect. 5.1 and Fig. 17). Moreover, in this region of potentials the "photocurrent doubling", associated with the injection of electrons into the conduction band, is expected to be slowed down by the small band bending. On the other hand, the multiplication of the photocurrent in the saturation region, especially when associated with a quantum efficiency higher than 1, is certainly indicative of the kind of mechanism described by Eqs. (47) and (48). A perfect "photocurrent doubling agent" has to combine the ability to compete very efficiently for the photogenerated holes with other reducing species (including water and OH^- ions) with the aptitude of the formed radical intermediates for interacting preferentially with the conduction band level. This kind of behaviour implies that both the original reducing species and the resulting radical intermediate undergo adsorption at the electrode surface.

6 Concluding Remarks

One of the specificities of the photoelectrochemical processes, involving minority charge carriers photogenerated in the semiconductor electrode, is that the current-potential relationship is of rather limited usefulness with regard to the determination of reaction kinetics. This is connected with the fact that, under such conditions, changes in electrode potential affect mainly the width of the space charge layer in the semiconductor (cf. Eq. (41)) and the density of the majority charge carriers at its surface (cf. Eq. (11)).

There is, however, a region of potentials, close to the flat-band potential, in which the amount of the observed photocurrent is controlled principally by the competition between the surface hole-electron recombination and (in the case of an n-type semiconductor) the hole transfer to the species on the solution side of the interface. Therefore, this initial portion of the photocurrent-potential curves enables a direct comparison of the reaction rates using, for example, a series of solutions containing variable amounts of two different reducing species (in practice, in the case of TiO$_2$ photoelectrode immersed in an aqueous solution, water will be always one of the species competing for the photo-generated holes). Actually, more rapid is a given photoanodic reaction, steeper is the departure of the photocurrent-potential curve and its onset potential is closer to the flat-band potential of the semiconducting electrode.

As the increase of light intensity tends to favour the surface recombination, due to larger availability of the photogenerated positive holes, this provides a possibility to render more apparent the differences in the photo-oxidation rates between various reducing species.

The analysis of the initial part of the photocurrent-potential curves may also yield useful indications regarding competitive photo-oxidation reactions occurring in photo-catalytic systems based on semiconductor powder suspensions. A special kind of the latter systems – colloidal semiconductor dispersion, are well suited for the application of different photolysis techniques, including pulsed laser techniques, which are an important source of kinetic data for numerous species, the formation of which can be monitored by a suitable spectroscopic method[236–238]. The only disadvantage, peculiar to such photolysis techniques, is that thus obtained kinetic data refer to a certain (in general, unknown) value of the photopotential, attained under particular, and not always reproducible, illumination conditions. This is not the case for the electrochemical measurements which can provide the necessary insight into the photoreactions occurring in the region of potentials around the flat-band potential of the semiconductor.

When referring to TiO$_2$-based photocatalytic systems it is important to note that, in most cases, the semiconducting oxide is associated there with a noble metal or/and a noble metal oxide catalyst. While the role played by these catalysts in (partial) cathodic reactions seems relatively well understood[239–242], it remains less clear with regard to the photoanodic reactions. In particular, the exact function of the extensively used ruthenium dioxide catalyst has been questioned[243]. The role of RuO$_2$ as a "hole-transfer" catalyst has, for example, been established through laser-photolysis kinetic studies in the case of photo-oxidation of halide (Br$^-$ and Cl$^-$) ions in colloidal titanium dioxide dispersions. In fact, the yields of Br$_2^-$ and Cl$_2^-$ radical anions, photogenerated in the course of these reactions,

$$X^- \quad + \quad h^+ \xrightarrow{\;X^-\;} X_2^- \; (X^- = Cl^-, Br^-) \tag{50}$$

were visibly enhanced in the presence of RuO$_2$ deposited onto TiO$_2$ particles[236]. On the other hand, Sakata et al.[243] have concluded, on the basis of a series of photolysis experiments with aqueous suspensions of RuO$_2$-loaded TiO$_2$ powders, that ruthenium dioxide acts in reality as a reduction catalyst. Their main argument was that the presence of RuO$_2$ caused a decrease of the rate of oxygen evolution during the photolysis of an AgNO$_3$ solution[243]:

$$4\,Ag^+ + 6\,H_2O \xrightarrow{\text{hv}} 4\,Ag + 4\,H_3O^+ + O_2 \tag{51}$$

However, the choice of the above reaction, involving the concomitant reduction of strongly oxidizing Ag^+ species, appears rather inappropriate for such a demonstration. As a matter of fact, given the positive value of the standard redox potential of the Ag^+/Ag couple (0.799 V), the mean photopotential of TiO_2 particles irradiated in an $AgNO_3$ solution may be expected to be sufficiently positive to ensure the anodic overvoltage, required by the slow step in the water photo-oxidation reaction (cf. Sect. 4.4), and to restrict the surface hole-electron recombination associated with the reduction of the peroxo-titanate species. Under such conditions, the conduction-band process, i.e., the deposition of metallic silver particles may eventually become rate-determining. The latter situation is therefore quite different from that characterizing the photolysis of water, with the photopotential of TiO_2 particles close to the flat-band potential and a particularly effective surface recombination.

There is no contradiction between the observations that ruthenium dioxide is effective in catalysing otherwise blocked photogeneration of oxygen, at potentials close to the flat-band potential of TiO_2 particles[18], but becomes ineffective at significant anodic overvoltages where the quantum efficiency for oxygen evolution, at TiO_2 photocatalyst immersed in an $AgNO_3$ solution, attains 25 per cent[243].

Finally, the most of inconsistencies and controversies regarding the activity of RuO_2 catalysts in various photocatalytic systems may certainly be explained by important differences in the methods used for their preparation. It is, in fact, well known that the electrochemical behaviour of ruthenium dioxide and, in particular, its anodic stability is influenced by the degree of hydration of the lattice[244]. The latter has probably little (if any) effect upon the initial activity of the $RuO_2 \cdot xH_2O$ catalyst (as measured, for example, by flash photolysis techniques) but has been shown to strongly affect its long-term ability to promote oxygen evolution[245]. The pretreatment of hydrated ruthenium dioxide at about 420 K has been reported[245] to ensure an optimum catalytic activity for the generation of oxygen and to suppress practically the competing oxidation reaction leading to the conversion of RuO_2 into gaseous RuO_4. A more global approach to the question of "photoelectrocatalysis" has recently been attempted by Contractor and Bockris[246] who have compared the effect of a large series of metals, deposited onto TiO_2 photoanodes, upon the photocurrent-voltage curves for oxygen evolution. Certain among the metals studied, in particular gold, palladium and iridium, have been shown to cause a significant increase of the anodic photocurrents measured in alkaline solution (1 M aqueous KOH) whilst only small changes were observed in acidic medium (0.5 M H_2SO_4)[246]. The amount of the cathodic shift of the photocurrent-potential curves, taken as a measure of the electrocatalytic effect of a given metal, has been found to increase with decreasing the bond energy D(M–OH). This correlation has been interpreted[246] in terms of the rate-determining step in oxygen photogeneration at TiO_2 involving metal-catalysed OH desorption.

More detailed studies are required in order to check whether the above correlation, established on the basis of relatively rapid potential sweep measurements, holds also for the steady-state photocurrents, i.e., in the situation when the TiO_2 surface becomes covered with the peroxo-titanate species. These should also include water photocleavage experiments onto titanium dioxide powders loaded with some of the catalysts investigated by Contractor and Bockris. The difficulty, associated with the fact that most of these metals are also hydrogen evolution catalysts, may be overcome, for example, by loading TiO_2 with both platinum and gold. The effectiveness of Pt in promoting the photogeneration of H_2 at TiO_2 is well established whilst Au, designated as the most active among metals catalysing the photo-oxidation of water[246], exhibits higher hydrogen over-voltage than Pt and, therefore, may be expected to affect less the cathodic process.

In this connection, it should be recalled that in the (close) water photocleavage systems, consisting of aqueous suspensions of TiO_2 powders loaded, e.g., with platinum or rhodium, no (or only negligible) oxygen evolution is observed (cf. Sect. 4). The photoanodic process remains blocked at the stage of peroxo-titanates and this limits the amount of photogenerated hydrogen and imposes a periodic regeneration of the photo-catalyst[135]. Now, if the activity of gold in catalysing the photoproduction of oxygen, i.e., in decreasing the overvoltage required for the O_2 evolution, extends to the potentials close to the flat-band potential, loading of the TiO_2 powder with the Au catalyst should result in the formation of molecular oxygen in such a system.

In the light of the above discussion (cf. Sects. 4 and 5), the formation of the surface-bonded peroxo species appears as the principal feature controlling the photoelectrochemical behaviour of n-type TiO_2:

- The photo-oxidation of the peroxo species constitutes most likely the slow step in the sequence of reactions leading to the formation of molecular oxygen (cf. Sect. 4.4).
- The ability of the surface-bonded peroxo species to undergo reduction, via conduction-band electrons, in the region of potentials close to that of reversible hydrogen electrode in the same solution, gives rise to very effective hole-electron recombination, decreasing the yield of photocatalytic dissociation of water.
- The latter explains also why a number of ions and molecules undergo photo-oxidation much easier than water and OH^- anions, especially at potentials close to the flat-band potential of TiO_2. The ability of such species to react directly with the precursors to the peroxo-titanates, i.e., with the $Ti_s-O^{\cdot}$ radicals or positive holes, permits not only to avoid the slow step in the process of photo-oxidation, but also to restrict the activity of the recombination centres by reducing the surface concentration of the peroxides.

It is, of course, tempting to draw a parallel between the photo-oxidation of water at titanium dioxide and the well-known irreversibility of O_2 generation at the anodes with metallic conductivity. There are, however, some indications that the pathway involving surface-bonded peroxo-titanate intermediates is not the only possible one. Recent experiments, performed with beryllium-doped TiO_2 photoanodes, have shown no evidence for the formation of any perceptible amount of the peroxo species[247]. This coincided with a substantial cathodic shift of the onset potential and a strong increase of the anodic photocurrent for these electrodes[247], suggesting a different mechanism of oxygen evolution.

Acknowledgments. I would like to thank Drs. Martine Ulmann and Alain Monnier for their help in reviewing the literature and for the critical reading of the manuscript.

This work was supported by the Swiss National Science Foundation; the grant from the E. and L. Schmidheiny Foundation is gratefully acknowledged. I am thankful to Mr. R. Cros for executing the drawings.

7 References

1. Renz, C.: Helv. Chim. Acta 4, 961 (1921)
2. Gion, L.: C.R. Acad. Sci. Paris 195, 421 (1932)
3. Jacobsen, A. E.: Ind. Eng. Chem. 41, 523 (1949)
4. Fujishima, A., Honda, K., Kikuchi, S.: J. Chem. Soc. Japan 72, 108 (1969)
5. Fujishima, A., Honda, K.: Nature 238, 37 (1972)
6. Augustynski, J., Hinden, J., Stalder, C.: J. Electrochem. Soc. 124, 1063 (1977)
7. Gosh, A. K., Maruska, H. P.: ibid. 124, 1516 (1977)
8. Houlihan, J. F., Armitage, D. B., Hoovler, T., Bonaquist, D., Mullay, L. N.: Mat. Res. Bull. 13, 1205 (1978)
9. Stalder, C., Augustynski, J.: J. Electrochem. Soc. 126, 2007 (1979)
10. Monnier, A., Augustynski, J.: ibid. 127, 1576 (1980)
11. Matsumoto, Y., Kurimoto, J., Shimizu, T., Sato, E.: ibid. 128, 1090 (1981)
12. Rajeshwar, K.: J. Appl. Electrochem. 15, 1 (1985)
13. Houlding, U. H., Grätzel, M.: J. Am. Chem. Soc. 105, 5695 (1983)
14. Desilvestro, J., Grätzel, M., Kavan, L., Moser, J., Augustynski, J.: ibid. 107, 2988 (1985)
15. Bulatov, A. V., Khidekel, M. L.: Izv. Akad. Nauk SSSR Ser. Khim. 1902 (1976)
16. Wagner, F. T., Somorjai, G. A.: Nature 285, 559 (1980)
17. Lehn, J. M., Sauvage, J. P., Ziessel, R.: Nouv. J. Chim. 4, 623 (1980)
18. Borgarello, E., Kiwi, J., Pelizzetti, E., Visca, M., Grätzel, M.: J. Am. Chem. Soc. 103, 6324 (1981)
19. Mills, A., Porter, G.: J. Chem. Soc., Faraday Trans. I 78, 3659 (1982)
20. Yesodharan, E., Grätzel, M.: Helv. Chim. Acta 66, 2145 (1983)
21. Kalyanasundaram, K., Grätzel, M., Pelizzetti, E.: Coord. Chem. Rev.
22. Pruden, A. L., Ollis, D. F.: J. Catal. 82, 404 (1983)
23. Pruden, A. L., Ollis, D. F.: Environ. Sci. Technol. 17, 628 (1983)
24. Ollis, D. F., Hsiao, C. Y., Budiman, L., Lee, C. L.: J. Catal. 88, 89 (1984)
25. Matthews, R. W.: J. Chem. Soc. Faraday Trans. I 80, 457 (1984)
26. Augugliaro, V., Lauricella, A., Rizzuti, L., Schiavello, M., Sclafani, A.: Int. J. Hydrogen Energy 7, 845 (1982)
27. Augugliaro, V., D'Alba, F., Rizzuti, L., Schiavello, M., Sclafani, A.: ibid. 7, 851 (1982)
28. Radford, P. P., Francis, C. G.: J. Chem. Soc. Chem. Commun. 1520 (1983)
29. Yue, P. L., Khan, F., Rizzuti, L.: Chem. Eng. Sci. 38, 1893 (1983)
30. Schrauzer, G. N., Guth, T. D., Salehi, J., Strampach, N., Liu Nan Hui, Palmer, N. R.: In Homogeneous and Heterogeneous Photocatalysis, NATO ASI Series, Series C, vol. 174 (Pelizzetti, E., Serpone, N. Eds.), Reidel, D., Dordrecht 1986, p. 509
31. Sato, S., White, J. M.: J. Am. Chem. Soc. 102, 7206 (1980)
32. Tsai, S. C., Chung, Y. W.: J. Catal. 86, 231 (1984)
33. Boehm, H. P.: Adv. Catalysis 16, 249 (1966)
34. Boehm, H. P.: Discuss. Faraday Soc. 52, 264 (1971)
35. Jackson, P., Parfitt, G. D.: Trans. Faraday Soc. 67, 2469 (1971)
36. Waldsax, J. C. R., Jaycock, M. J.: Discuss. Faraday Soc. 52, 231 (1971)
37. Jaycock, M. J., Waldsax, J. C. R.: J. Chem. Soc. Faraday Trans. I 70, 1501 (1974)
38. Boehm, H. P., Herrmann, M.: Z. Anorg. Chem. 352, 156 (1967)
39. Munuera, G., Stone, F. S.: Discuss. Faraday Soc. 52, 205 (1971)
40. Gonzalez, F., Munuera, G.: Rev. Chim. Minérale 7, 1021 (1970)
41. Munuera, G., Rives-Arnau, V., Saucedo, A.: J. Chem. Soc. Faraday Trans. I 75, 736 (1979)
42. Yates, D. J. C.: J. Phys. Chem. 65, 746 (1961)
43. Primet, M., Pichat, P., Mathieu, M. V.: C.R. Acad. Sci. Ser. B 267, 799 (1968)
44. Primet, M., Pichat, P., Mathieu, M. V.: J. Phys. Chem. 75, 1216 (1971)
45. Little, L. H.: in Infrared Spectra of Adsorbed Species, Academic Press, London and New York, 1966, chap. 10
46. Jones, P., Hockey, J. A.: Trans. Faraday Soc. 67, 2669 (1971)
47. Sham, T. K., Lazarus, M. S.: Chem. Phys. Lett. 68, 426 (1979)
48. Lo, W. J., Chung, Y. W., Somorjai, G. A.: Surf. Sci. 71, 199 (1978)

49. Lewis, K. E., Parfitt, G. D.: Trans. Faraday Soc. *62*, 204 (1966)
50. Johnson, O. W., Ohlsen, W. D., Kingsbury, P. I.: Phys. Rev. *175*, 1102 (1968)
51. Jones, P., Hockey, J. A.: J. Chem. Soc. Faraday Trans. I *68*, 907 (1972)
52. Bates, J. B., Perkins, R. A.: Phys. Rev. B *16*, 3713 (1977)
53. Jones, P., Hockey, J. A.: Trans. Faraday Soc. *67*, 2679 (1971)
54. Enriquez, M. A., Dorémieux-Morin, C., Fraissard, J.: J. Solid State Chem. *40*, 233 (1981)
55. Ilier, R. K.: in The Colloid Chemistry of Silica and Silicates, Cornell Univ. Press. Ithaca, N.Y. 1955
56. Tadros, Th. F., Lyklema, J.: J. Electroanal. Chem. Interfacial Electrochem. *17*, 267 (1968)
57. Tadros, Th. F., Lyklema, J.: ibid. *22*, 1 (1969)
58. Ganichenko, L. G., Kiselev, V. F., Murina, V. V.: Kin. i Kat. *2*, 877 (1961)
59. Iwaki, T., Miura, M.: Bull. Chem. Soc. Jpn. *44*, 1754 (1971)
60. Parfitt, G. D.: Prog. Surf. Membr. Sci. *11*, 181 (1976)
61. Misra, D. N.: Nature (Phys. Sci.) *240*, 14 (1972)
62. Boehm, H. P.: Angew. Chem. Int. Edn *5*, 541 (1966)
63. Lyklema, J.: Discuss. Faraday Soc. *52*, 276 (1971)
64. Parks, G. A.: Chem. Rev. *65*, 177 (1965)
65. Everett, D. H. (Ed.): in Definitions, Terminology and Symbols in Colloid and Surface Chemistry. Part 1. Butterworths, London 1972
66. Levine, S., Smith, A. L.: Discuss. Faraday Soc. *52*, 290 (1971)
67. Breeuwsma, A., Lyklema, J.: ibid. *52*, 324 (1971)
68. Payman, M. A. F., Bowden, J. W., Posner, A. M.: Aust. J. Soil Res. *17*, 191 (1979)
69. Furlong, D. N., Yates, D. E., Healy, T. W.: in Electrodes of Conductive Metallic Oxides. Part B. (Trasatti, S. Ed.), Elsevier, Amsterdam 1981, chap. 8
70. Yates, D. E., Healy, T. W.: J. Colloid Interface Sci. *35*, 9 (1976)
71. Onoda, G. Y., de Bruyn, P. L.: Surf. Sci. *4*, 48 (1966)
72. Bérubé, Y. G., Onoda Jr., G. Y., de Bruyn, P. L.: ibid. *8*, 448 (1967)
73. Bérubé, Y. G., de Bruyn, P. L.: J. Colloid Interface Sci. *27*, 305 (1968)
74. Blok, L., de Bruyn, P. L.: ibid. *32*, 518, 527, 544 (1970)
75. Yates, D. E., Healy, T. W.: J. Chem. Soc. Faraday Trans. I *76*, 9 (1980)
76. Ahmed, S. M.: in Oxide and Oxides Films, Vol. 1 (Diggle, J. W. Ed.), Marcel Dekker, New York 1972, p. 230
77. Wiese, G. R., Healy, T. W.: J. Colloid Interface Sci. *51*, 434 (1975)
78. Bobyrenko, Yu. Ya., Zholnin, A. B., Konovalova, V. K.: Russ. J. Phys. Chem. *46*, 749 (1972)
79. Schindler, P. W., Gamsjäger, H.: Kolloid Z. and Polymere *250*, 759 (1972)
80. Johansen, P. G., Buchanan, A. S.: Aust. J. Chem. *10*, 398 (1957)
81. Fukuda, H., Miura, M.: J. Sci. Hiroshima Univ. Ser. A *36*, 77 (1972)
82. Iwaki, T., Komuro, M., Miura, M.: Bull. Chem. Soc. Jpn. *45*, 2343 (1972)
83. Bérubé, Y. G., de Bruyn, P. L.: J. Colloid Interface Sci. *28*, 92 (1968)
84. Yates, D. E., Levine, S., Healy, T. W.: J. Chem. Soc. Faraday Trans. I *70*, 1807 (1977)
85. Healy, T. W., White, L. R.: Adv. Colloid Interface Sci. *9*, 303 (1978)
86. Bockris, J. O'M., Anderson, T. N.: Electrochim. Acta *9*, 347 (1964)
87. Davis, J. A., James, R. O., Leckie, J. O.: J. Colloid Interface Sci. *63*, 480 (1978)
88. James, R. O., Leckie, J. O., Davis, J. A.: ibid. *65*, 331 (1978)
89. Lyklema, J.: J. Electroanal. Chem. Interfacial Electrochem. *18*, 341 (1968)
90. Levine, S., Smith, A. L., Brett, A. C.: in VI International Congress of Surface-active Agents, Zürich 1972. Hanser, München 1972, p. 603
91. Perram, J. W.: J. Chem. Soc. Faraday Trans. II *69*, 993 (1973)
92. Perram, J. W., Hunter, R. J., Wright, H. J. L.: Chem. Phys. Lett. *23*, 265 (1973)
93. Perram, J. W., Hunter, R. J., Wright, H. J. L.: Aust. J. Chem. *27*, 461 (1974)
94. Grahame, D. C.: Chem. Rev. *41*, 441 (1947)
95. Lyklema, J.: Trans. Faraday Soc. *59*, 418 (1963)
96. Dumont, F., Watillon, A.: Discuss. Faraday Soc. *52*, 352 (1971)
97. Gurney, R. W.: in Ionic Processes in Solutions, Dover, New York 1953
98. Conway, B. E.: in Ionic Hydration in Chemistry and Biophysics, Elsevier, Amsterdam 1981, p. 641
99. Barclay, D. J.: J. Electroanal. Chem. Interfacial Electrochem. *19*, 318 (1968)

100. Conway, B. E., Angerstein-Kozlowska, H., Sharp, W. B. A.: Z. Phys. Chem. N.F. *98*, 61 (1975)
101. Conway, B. E.: in Ionic Hydration in Chemistry and Biophysics, Elsevier, Amsterdam 1981, p. 614
102. Sanchez, J., Augustynski, J.: J. Electroanal. Chem. Interfacial Electrochem. *103*, 423 (1979)
103. Augustynski, J., Koudelka, M., Sanchez, J.: in the Extended Abstracts of the 31st Meeting of the International Society of Electrochemistry, Venice, Sept. 22–26, 1980, p. 208
104. Koudelka, M., Monnier, A., Augustynski, J.: J. Electrochem. Soc. *131*, 745 (1984)
105. Boehm, H. P.: Discuss. Faraday Soc. *52*, 164 (1971)
106. Malati, M. A., Seager, N. J.: J. Oil Colour Chem. Assoc. *64*, 231 (1981)
107. Petit, J., Poisson, R.: C.R. Acad. Sci. Paris *240*, 312 (1955)
108. Flaig-Baumann, R., Herrmann, M., Boehm, H. P.: Z. Anorg. Chem. *372*, 296 (1970)
109. Herrmann, M., Kaluza, U., Boehm, H. P.: ibid. *372*, 308 (1970)
110. Munuera, G., Navio, J. A., Rives-Arnaud, V.: in Fourth International Conference on Photochemical Conversion and Storage of Solar Energy (Rabbani, J. Ed.), Jerusalem 1982, p. 141
111. Munuera, G., Gonzalez-Elipe, A. R., Rives-Arnaud, V., Navio, A., Malet, P., Soria, J., Conesa, J. C., Sanz, J.: in Studies in Surface Science and Catalysis (Che, M., Bond, G. C. Eds.), Elsevier, Amsterdam 1985, vol. 21, p. 113
112. Koudelka, M., Sanchez, J., Augustynski, J.: J. Phys. Chem. *86*, 4277 (1982)
113. Takenaka, Y., Nakatani, M., Sugimori, S., Uchida, H.: Nippon Kagaku Kaishi *9*, 1650 (1985)
114. Sanchez, J., Augustynski, J.: unpublished results
115. Augustynski, J., Berthou, H., Painot, J.: Chem. Phys. Lett. *44*, 221 (1975)
116. Augustynski, J.: in Passivity of Metals (Frankenthal, R. P., Kruger, J. Eds.). The Electrochemical Soc., Inc., Princeton 1978, p. 989
117. Kazarinov, V. E., Andreev, V. N., Mayorov, A. P.: J. Electroanal. Chem. Interfacial Electrochem. *130*, 277 (1981)
118. Morrison, R. S.: in Electrochemistry at Semiconductor and Oxidized Metal Electrodes, Plenum Press, New York and London, 1980
119. Dutoit, E. C., Cardon, F., Vanden Kerchove, F., Gomes, W. P.: J. Appl. Electrochem. *8*, 247 (1978)
120. Schmickler, W., Schultze, J. W.: in Modern Aspects of Electrochemistry (Bockris, J. O'M., Conway, B. E., White, E. Eds.), Plenum Press, N.Y. 1986, vol. 17, chap. 5
121. Vanden Berghe, R. A. L., Cardon, F., Gomes, W. P.: Surf. Sci. *39*, 368 (1973)
122. Schrauzer, G. N., Guth, T. D.: J. Am. Chem. Soc. *99*, 7189 (1977)
123. van Damme, H., Hall, W. K.: ibid. *101*, 4373 (1979)
124. Domen, K., Naito, S., Sorna, M., Onishi, T., Tamaru, K.: J. Chem. Soc., Chem. Commun. 543 (1980)
125. Sato, S., White, J. M.: Chem. Phys. Lett. *72*, 83 (1980)
126. Kawai, T., Sakata, T.: ibid. *72*, 87 (1980)
127. Borgarello, E., Kiwi, J., Pelizzetti, E., Visca, M., Grätzel, M.: Nature (London) *289*, 158 (1981)
128. Grätzel, M.: Acc. Chem. Res. *14*, 376 (1981)
129. Grätzel, M. (Ed.): in Energy Resources through Photochemistry and Catalysis, Academic Press, New York 1983
130. Pichat, P.: in Photoelectrochemistry, Photocatalysis and Photoreactors, NATO ASI Series, Series C, Vol. 146 (Schiavello, M. Ed.), Reidel D., Dordrecht 1985, p. 425
131. Heller, A.: Science *223*, 1141 (1984)
132. Albery, W. J., Bartlett, P. N.: J. Electrochem. Soc. *131*, 315 (1984)
133. Yesodharan, E., Yesodharan, S., Grätzel, M.: Sol. Energy Mater. *10*, 287 (1984)
134. Duonghong, D., Grätzel, M.: J. Chem. Soc., Chem. Commun. 1597 (1984)
135. Gu, B., Kiwi, J., Grätzel, M.: Nouv. J. Chim. *9*, 539 (1985)
136. Sawyer, D. T., Valentine, J. S.: Acc. Chem. Res. *14*, 393 (1981)
137. Ulmann, M., de Tacconi, N., Augustynski, J.: J. Phys. Chem. *90*, 6523 (1986)
138. Wilson, R. H.: J. Electrochem. Soc. *127*, 228 (1980)
139. Salvador, P., Gutiérrez, C.: Chem. Phys. Lett. *86*, 131 (1982)
140. Salvador, P., Gutiérrez, C.: J. Phys. Chem. *88*, 3696 (1984)
141. Koudelka, M., Monnier, A., Sanchez, J., Augustynski, J.: J. Mol. Catal. *25*, 295 (1984)

142. Monnier, A., Augustynski, J., Stalder, C.: J. Electroanal. Chem. Interfacial Electrochem. *112*, 383 (1980)
143. Conway, B. E.: in Electrodes of Conductive Metallic Oxides. Part B (Trasatti, S. Ed.), Elsevier, Amsterdam 1981, p. 439
144. Hoare, J. P.: in Encyclopedia of Electrochemistry of the Elements (Bard, A. J. Ed.), Dekker, M., New York 1974, vol. II, chap. 5
145. Boonstra, A. H., Mutsaers, C. A. H. A.: J. Phys. Chem. *79*, 1940 (1975)
146. Pappas, S. P., Fischer, R. M.: J. Paint Techn. *46*, 65 (1974)
147. Rao, M. V., Rajeshwar, K., Pai Verneker, V. R., DuBow, J.: J. Phys. Chem. *84*, 1987 (1980)
148. Salvador, P.: J. Electrochem. Soc. *128*, 1895 (1981)
149. Salvador, P.,Decker, F.: J. Phys. Chem. *88*, 6116 (1984)
150. Baur, E.: Helv. Chim. Acta *1*, 186 (1917)
151. Baur, E., Neuweiler, C.: ibid. *10*, 901 (1927)
152. Winter, G.: Nature (London) *163*, 326 (1949)
153. Markham, M. C., Laidler, K. J.: J. Phys. Chem. *57*, 363 (1953)
154. Rubin, T. R., Calvert, J. G., Rankin, G. T., MacNevin, W. M.: J. Am. Chem. Soc. *75*, 2850 (1953)
155. Calvert, J. G., Theurer, K., Rankin, G. T., MacNevin, W. M.: ibid. *76*, 2575 (1954)
156. Dixon, D. R., Healy, T. W.: Aust. J. Chem. *24*, 1193 (1971)
157. Irick, G.: J. Appl. Polym. Sci. *16*, 2387 (1972)
158. Harbour, J. R., Hair, M. L.: J. Phys. Chem. *83*, 652 (1979)
159. Hauffe, R.: Rev. Pure Appl. Chem. *18*, 79 (1968)
160. Harbour, J. R., Tromp, J., Hair, M. L.: Can. J. Chem. *63*, 204 (1985)
161. Harbour, J. R., Hair, M. L.: J. Phys. Chem. *81*, 1791 (1977)
162. Wrighton, M. S., Ginley, D. S., Wolczanski, D. T., Ellis, A. B., Morse, D. L., Linz, A.: Proc. Nat. Acad. Sci. USA *72*, 1518 (1975)
163. Kung, H. H., Jarrett, H. S., Sleight, A. W., Ferretti, A.: J. Appl. Phys. *48*, 2463 (1977)
164. Parkinson, B., Decker, F., Juliao, J. F., Abramovich, M., Chagas, H. C.: Electrochim. Acta *25*, 521 (1980)
165. Kobayashi, T., Yoneyama, H., Tamura, H.: J. Electrochem. Soc. *130*, 1706 (1983)
166. Damjanovic, A., Genshaw, M. A., Bockris, J. O'M.: ibid. *114*, 466 (1967)
167. Damjanovic, A.: in Modern Aspects of Electrochemistry (Bockris, J. O'M., Conway, B. E. Eds.), Plenum Press, N.Y. 1969, vol. 5, chap. 5
168. Gerischer, H.: J. Electroanal. Chem. Interfacial Electrochem. *150*, 553 (1983)
169. Brown, G. T., Darwent, J. R.: J. Phys. Chem. *88*, 4955 (1984)
170. Ferrer, I. J., Muraki, H., Salvador, P.: ibid. *90*, 2805 (1986)
171. Kautek, W., Gerischer, H.: Electrochim. Acta *26*, 1774 (1981)
172. Baxendale, J. H., Wilson, J. A.: Trans. Faraday Soc. *53*, 344 (1957)
173. Jaeger, C. D., Bard, A. J.: J. Phys. Chem. *83*, 3146 (1979)
174. Howe, R. F., Grätzel, M.: ibid. *89*, 4495 (1985)
175. Anpo, M., Shima, T., Kubokawa, Y.: Chem. Lett. 1799 (1985)
176. Howe, R. F., Grätzel, M.: J. Phys. Chem. *91*, 3906 (1987)
177. Völz, H. G., Kämpf, G., Fitzky, H. G.: Farbe u. Lack *78*, 1037 (1972)
178. Nikisha, V. V., Shelimov, B. N., Kazanskii, V. B.: Kinet. Katal. *12*, 332 (1971)
179. Nikisha, V. V., Shelimov, B. N., Kazanskii, V. B.: ibid. *15*, 676 (1974)
180. Meriaudeau, P., Vedrine, J. C.: J. Chem. Soc. Faraday Trans. II *72*, 472 (1976)
181. Gonzalez-Elipe, A. R., Munuera, G., Soria, J.: J. Chem. Soc. Faraday Trans. I *75*, 748 (1979)
182. Anpo, M., Aikawa, N., Kubokawa, Y., Che, M., Louis, C., Giamello, C.: J. Phys. Chem. *89*, 5689 (1985)
183. Anpo, M., Aikawa, N., Kubokawa, Y.: J. Chem. Soc. Chem. Commun. 644 (1984)
184. Naccache, C., Meriaudeau, P., Che, M., Tench, A. J.: Trans. Faraday Soc. *67*, 506 (1971)
185. Serwicka, E., Schlierkamp, M. W., Schindler, R. N.: Z. Naturforsch. *36a*, 226 (1981)
186. Rothenberger, G., Moser, J., Grätzel, M., Serpone, N., Sharma, D. K.: J. Am. Chem. Soc. *107*, 8054 (1985)
187. Kiwi, J., Grätzel, M.: J. Molecular Catal. *39*, 63 (1987)
188. Oosawa, Y., Grätzel, M.: J. Chem. Soc., Chem. Commun. 1629 (1984)

189. Bickley, R. I., Jayanty, R. K. M., Vishwanathan, V.: in Homogeneous and Heterogeneous Photocatalysis, NATO ASI Series, Series C, vol. 174 (Pelizzetti, E., Serpone, N. Eds.), Reidel, D., Dordrecht 1986, p. 555
190. Nishimoto, S., Ohtani, B., Kagiwara, H., Kagiya, T.: J. Chem. Soc. Faraday Trans. I 79, 2685 (1983)
191. Ulmann, M., Augustynski, J.: Chem. Phys. Lett. 141, 154 (1987)
192. Thompson, R. C.: Inorg. Chem. 23, 1794 (1984)
193. Hoare, J. P.: J. Electrochem. Soc. 112, 1129 (1965)
194. Harris, L. A., Wilson, R. H.: ibid. 123, 1010 (1976)
195. Harris, L. A., Cross, D. R., Gerstner, M. E.: ibid. 124, 839 (1977)
196. Gärtner, W. W.: Phys. Rev. 116, 84 (1959)
197. Butler, M. A.: J. Appl. Phys. 48, 1914 (1977)
198. Augustynski, J.: J. Electroanal. Chem. Interfacial Electrochem. 145, 457 (1983)
199. Bickley, R. I., Munuera, G., Stone, F. S.: J. Catal. 31, 398 (1973)
200. Frank, S. N., Bard, A. J.: J. Am. Chem. Soc. 99, 4667 (1977)
201. Inoue, T., Watanabe, T., Fujishima, A., Honda, K.: Chem. Lett. 1073 (1977)
202. Fujishima, A., Inoue, T., Watanabe, T., Honda, K.: ibid. 357 (1978)
203. Fujishima, A., Inoue, T., Honda, K.: J. Am. Chem. Soc. 101, 5582 (1979)
204. Dutoit, E. C., Cardon, F., Gomes, W. P.: Ber. Bunsenges. Phys. Chem. 80, 1285 (1976)
205. Kobayashi, T., Yoneyama, H., Tamura, H.: J. Electroanal. Chem. Interfacial Electrochem. 122, 133 (1981)
206. Grätzel, M.: Personnal communication
207. Hale, J. M.: in Reactions of Molecules at Electrodes (Hush, N. S. Ed.), Wiley Interscience, London 1971
208. Memming, R., Möllers, F.: Ber. Bunsenges. Phys. Chem. 76, 475 (1972)
209. Marcus, R. A.: J. Phys. Chem. 67, 853 (1963)
210. Marcus, R. A.: J. Chem. Phys. 43, 679 (1965)
211. Gerischer, H.: Z. Phys. Chem. (Frankfurt) 26, 223 (1960)
212. Gerischer, H.: ibid. 27, 40 (1961)
213. Gerischer, H.: ibid. 27, 48 (1961)
214. Nakato, Y., Tsumura, A., Tsubomura, H.: J. Phys. Chem. 87, 2402 (1983)
215. Nakato, Y., Ogawa, H., Morita, K., Tsubomura, H.: ibid. 90, 6210 (1986)
216. Vetter, K. J.: in Electrochemical Kinetics, Academic Press, New York 1967
217. Memming, R., Möllers, F.: Ber. Bunsenges. Phys. Chem. 76, 609 (1972)
218. Morrison, S. R., Freund, T.: J. Chem. Phys. 47, 1543 (1967)
219. Gomes, W. P., Freund, T., Morrison, S. R.: J. Electrochem. Soc. 115, 818 (1968)
220. Morrison, S. R., Freund, T.: Electrochim. Acta 13, 1343 (1968)
221. Gomes, W. P., Cardon, F.: J. Solid State Chem. 3, 125 (1971)
222. Cardon, F., Gomes, W. P.: Surf. Sci. 27, 286 (1971)
223. Micka, K., Gerischer, H.: J. Electroanal. Chem. Interfacial Electrochem. 38, 397 (1972)
224. Gerischer, H., Rösler, H.: Chem. Ing. Tech. 42, 176 (1970)
225. Vanden Berghe, R. A. L., Gomes, W. P., Cardon, F.: Z. Phys. Chem. N.F. 92, 91 (1974)
226. Vanden Kerchove, F., Vandermolen, J., Gomes, W. P.: Ber. Bunsenges. Phys. Chem. 83, 230 (1979)
227. Yoneyama, H., Toyuguchi, Y., Tamura, H.: J. Phys. Chem. 76, 3460 (1972)
228. Benderskii, V. A., Zolotovitskii, Ya. M., Kogan, Ya. L., Khidekel', M. L., Shub, D. M.: Dokl. Akad. Nauk SSR 222, 606 (1975)
229. Miyake, M., Yoneyama, H., Tamura, H.: Chem. Lett. 635 (1976)
230. Miyake, M., Yoneyama, H., Tamura, H.: Electrochim. Acta 22, 319 (1977)
231. Maeda, Y., Fujishima, A., Honda, K.: J. Electrochem. Soc. 128, 1731 (1981)
232. Hirano, K., Inagaki, K., Asami, Y., Takagi, R.: J. Electrochem. Soc. Jpn. 51, 893 (1983)
233. Rose, T. L., Nanjundiah, C.: J. Phys. Chem. 89, 3766 (1985)
234. Hykaway, N., Sears, W. M., Morisaki, H., Morrison, S. R.: ibid. 90, 6663 (1986)
235. Miyake, M., Yoneyama, H., Tamura, H.: J. Catal. 58, 22 (1979)
236. Moser, J., Grätzel, M.: Helv. Chim. Acta 65, 1436 (1982)
237. Henglein, A.: Ber. Bunsenges. Phys. Chem. 86, 241 (1982)
238. Moser, J., Grätzel, M.: J. Am. Chem. Soc. 105, 6547 (1983)

239. Yamamoto, N., Tonomura, S., Matsuoka, T., Tsubomura, H.: Surf. Sci. *92,* 400 (1980)
240. Aspnes, D. E., Heller, H.: J. Phys. Chem. *87,* 4919 (1983)
241. Sakata, T., Kawai, T., Hashimoto, K.: Chem. Phys. Lett. *88,* 50 (1982)
242. Nakabayashi, S., Fujishima, A., Honda, K.: ibid. *102,* 464 (1983)
243. Sakata, T., Hashimoto, K., Kawai, T.: J. Phys. Chem. *88,* 5214 (1984)
244. Galizzioli, D., Tantardini, F., Trasatti, S.: J. Appl. Electrochem. *4,* 57 (1974)
245. Mills, A., Lawrence, C., Enos, R.: J. Chem. Soc. Chem. Commun. 1436 (1984)
246. Contractor, A. Q., Bockris, J. O'M.: Electrochim. Acta *32,* 121 (1987)
247. Ulmann, M., Augustynski, J.: to be published

Excited States of Chromium(III) in Translucent Glass-Ceramics as Prospective Laser Materials

Renata Reisfeld[1] and Christian K. Jørgensen[2]

1 Enrique Berman Professor of Solar Energy; Department of Inorganic and Analytical Chemistry, The Hebrew University of Jerusalem, Jerusalem 91904, Israel
2 Département de Chimie, Analytique et Appliquée, Université de Genève, CH 1211 Geneva 4, Switzerland

The first excited quartet state $^4T_2$ of chromium(III) provides broad-band luminescence in the near infrared. If the first doublet state ^2E is below the $^4T_2$ potential minimum, a narrow band may be emitted with high yield, as known from ruby. Glass-ceramics containing Cr(III) may show high quantum efficiency of $^4T_2$ at room temperature, contrary to most other compounds. When the heat treatment is such that micro-crystallites of spinel, gahnite, mullite, virgilite,... are obtained having dimensions smaller than the wave-length of violet light, transparent glassy-looking materials are obtained which may be suitable for tunable lasers similar to alexandrite, and luminescent solar concentrators. The absorption and emission spectra of Cr(III) are treated here in the angular overlap model. The sub-shell energy difference Δ (= 10 Dq) of Cr(III) in glasses is around 0.86 and in glass-ceramics 0.96 to 1.05 times its value of the aqua ion. Evidence is presented that the band widths and Stokes shift cannot be fully explained by effects of totally symmetric stretching. The multi-dimensional potential surfaces, combined with the Tanabe-Kamimura theorem and the Jahn-Teller effect, provide a more complete explanation of the spectroscopic behaviour. Anti-ferromagnetic and spin-orbit coupling are shortly reviewed. Comparison between chromium(III) luminescence and the exceptional Stokes shift of hexacyanocobaltate(III) is made.

Structure and Bonding 69
© Springer-Verlag Berlin Heidelberg 1988

1 Introduction

Chromium doped transparent glass-ceramics having microcrystallites of dimensions smaller than the wavelength of visible light are prepared by first melting the appropriate Cr^{3+} doped glasses at 1560 to 1580 °C and then carefully following the cooling curve of the glass. Provided enough time is given the glass is transformed from the non-equilibrium glassy state to crystalline phases in which the thermodynamical equilibrium is reached. Conditions for the preparation of five types of glass-ceramics, spinel-type $(MgAl_{2-x}Cr_xO_4)$[1], gahnite-type $(ZnAl_{2-x}Cr_xO_4)$[1,2], β-quartz-like[3], petalite-like $(SiO_2Al_2O_3RO$ where $R = Mg, Ca, Zn)$[3], mullite[4] and virgilite[5], containing Cr^{3+}, were extensively studied recently.

The importance of transparent glass-ceramics doped by Cr^{3+} as potential new materials for luminescent solar concentrators and tunable lasers is considerable[6,7]. It has been shown that Cr(III) exhibits exceptionally high quantum efficiency of luminescence in these materials as compared to glasses of the same compositions[1,3,6,8,9]. Cr^{3+} is also used as a probe for nucleation of and crystallization of cordierite into glass-ceramics[10].

2 Luminescent Solar Concentrators (LSC)

One of the ways of decreasing the price of photovoltaic (PV) electricity is by concentrating solar light onto small areas thus reducing the amount of expensive solar cells which convert light to electricity. This can be achieved by luminescent solar concentrators which absorb solar light on large plate areas and emit the concentrated fluorescent light from the small area of the plate edges to which PV cells are attached.

A plate of highly transparent material consists of a high quantum efficient luminescent species which absorbs the solar light. The resulting luminescence is re-emitted isotropically. Following Snell's law about 75–80% of the emitted photons are trapped within the plate by total internal reflection and transported to the edge of the plate where the concentrated fluorescent light impinges on PV cells.

The light is concentrated as a result of geometrical gain which is defined as the ratio of the exposed plate area to the escaping area.

The operating of an LSC is based on the absorption of solar radiation in a collector containing a fluorescent species in which the emission bands have little or no overlap with the absorption bands. The fluorescence emission is trapped by total internal reflection and concentrated at the edges of the collector which is usually a thin plate. LSC's have the following advantages over conventional solar concentrators: they collect both direct and diffuse light; there is good heat dissipation of non-utilized energy by the large area of the collector plate in contact with air so that essentially "cold light" reaches the PV cells; tracking the sun is unnecessary; the luminescent species can be chosen to allow matching of the concentrated light to the maximum sensitivity of the PV cells.

In our laboratory we have extensively studied inorganic ions in glasses as possible materials for laser and solar devices. The photostability of inorganic ions in glasses as well as in crystals is extremely high as evidenced by the variety of inorganic minerals found in nature. These materials preserved their optical properties for ages without any change under solar radiation and weather conditions.

The optical plate efficiency of the LSC depends on the following factors: (1) the fraction η_{abs} of the light absorbed; (2) the quantum efficiency η_l of the fluorescent species; (3) the Stokes efficiency η_s which is the ratio of the average energy of the emitted photons to the absorbed photon and is given by

$$\eta_s = \bar{\nu}_{emi}/\bar{\nu}_{abs} \tag{1}$$

(4) the fraction η_t of the light trapped in the collector which is given by

$$\eta_t = (1 - 1/n^2)^{1/2} \tag{2}$$

where n is the refractive index of the light-emitting medium; (5) the transport efficiency η_{tr} which takes the transport losses due to matrix absorption and scattering into account; (6) the efficiency η_{sel} due to losses arising from self-absorption of the colorants. The expression for the optical efficiency including reflection is given by

$$\eta_{opt} = (1 - R)\eta_{abs}\eta_l\eta_s\eta_{tr}\eta_{sel} \tag{3}$$

where R is the Fresnel reflection coefficient and is about 4%.

In view of their spectroscopic properties Cr^{3+}-doped glasses look very promising as materials for LSC's. The two broad absorption bands, $^4A_2 \rightarrow {}^4T_1$ which peaks at 450 nm and $^4A_2 \rightarrow {}^4T_2$ which peaks at 650 nm, cover the major part of the solar spectrum and the $^4T_2 \rightarrow {}^4A_2$ emission band peaking at around 850 nm matches the maximum sensitivity of the silicon solar cells. Inorganic oxide glasses doped with Cr^{3+} are extremely stable materials and can be easily prepared in large plates. The only drawback is that the quantum efficiency of the luminescence of Cr^{3+} in glasses is low, the highest value obtained so far being 0.17 in lithium calcium silicate glass and 0.22 in lithium lanthanum phosphate[11, 12]. These values, although much higher than for any $^4T_2 \rightarrow {}^4A_2$ emission of Cr^{3+} complexes, at room temperature are still too low for LSC's.

The relatively low efficiency of $^4T_2 \rightarrow {}^4A_2$ luminescence in glasses is also reflected by much smaller decay times than in crystals in which the quantum efficiences are almost 100% at room temperature. The reason for low quantum efficiencies in glasses arises from high nonradiative relaxation of the excited $^4T_2$ level which competes with radiative processes. The theory of nonradiative relaxation of Cr^{3+} in glasses is quite complicated[6].

We have found that the nonradiative relaxation can be decreased when Cr^{3+} is incorporated into glass-ceramics which result from crystallization of the glasses when exposed to special heat treatment. We have studied the conditions for such treatment in order to be able to tailor the specific glass-ceramics in question.

3 Requirements for Glass Lasers and Luminescent Solar Concentrators; Similarities and Differences

The requirements for a glass laser are:
a) high absorption of the exciting light
b) population inversion of the emitting level
c) high cross-section of stimulated laser emission
d) nonradiative quick relaxation of the lower laser level

The requirements of LSC's are:
a) large absorption in a broad spectral range
b) high population of the emitting level
c) high quantum efficiency of light emission
d) Stokes shift between the emitted and absorbed light

As can be seen the requirements for both devices are quite similar, however, they are more stringent for lasers than for LSC's. An additional need is that the materials will be stable towards corrosion by light, heat and humidity.

4 Vibronic Lasers

From the point of view of mechanical and optical properties vibronic lasers are connected to luminescent solar concentrators.

Vibronic lasers in which the terminating level in the laser transition is an excited coupled vibrational and electronic state include organic dyes in liquid solutions[13] and colour centre lasers[14]. On the basis of similar physical principles, a broadly tunable vibronic laser performance at room temperature has been demonstrated in alexandrite $(Cr^{3+}:BeAl_2O_4)$[15], Cr^{3+}-doped gadolinium scandium gallium garnets $(Cr^{3+}:Gd-Sc-Ga)$[16] and Cr^{3+}-doped gadolinium scandium aluminium garnet $(Cr^{3+}:Gd_3Sc_2Al_3O_{12})$[17]. The advantage of the solid state tunable laser arises from the long storage time of Cr^{3+} which in alexandrite is 260 µs, allowing significant energy storage and Q switch operation[6, 7]

Transparent glass-ceramics doped with Cr^{3+} may have optical properties comparable with alexandrite crystals. In both lasers and LSC's the excitation energy is absorbed in the shorter wavelength part of the spectrum and the resulting fluorescence is emitted at longer wavelengths. The Stokes shift of the emitted light arises from vibronic interactions of the excited and ground states of Cr^{3+}. Due to this shift self-absorption is minimal.

5 Spectroscopy of Chromium(III)

In the specific case of Cr^{3+} which we are dealing with here, the luminescence originates from transitions in the unfilled 3d electronic shell. Such transitions are Laporte-forbidden in sites having a centre of symmetry. The designations of the transitions in Cr^{3+} in

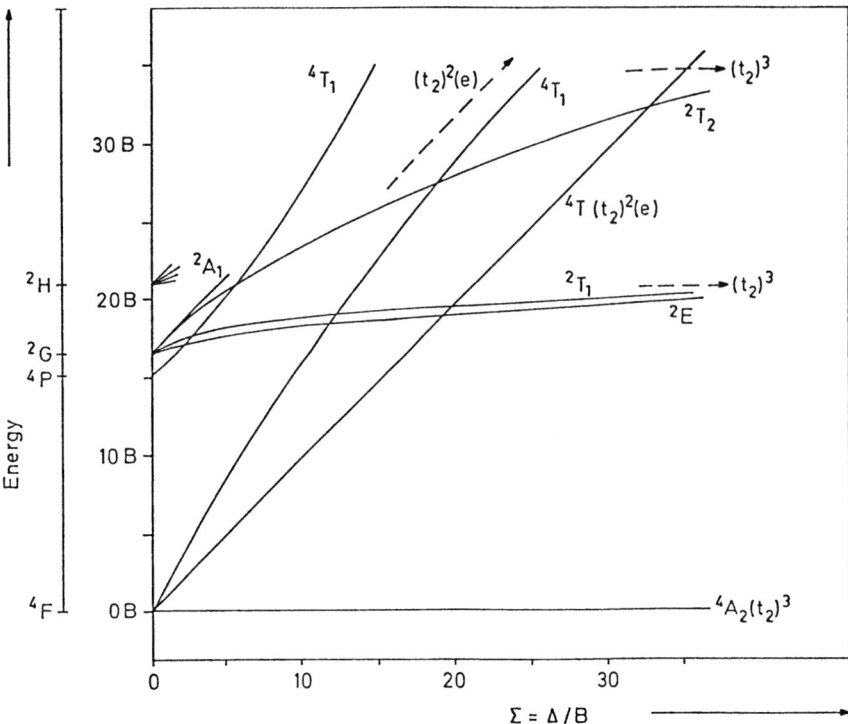

Fig. 1. Energy levels relative to the groundstate $^4A_2$ of octahedral d^3 systems (with negligible spin-orbit coupling) such as chromium(III). This Tanabe-Sugano diagram has the Racah parameter B of interelectronic repulsion as unit of energy. The variable is the ligand field strength $\Sigma = \Delta/B$, the ratio between the sub-shell energy difference Δ and B. The levels $^4A_2$, $^4T_2$ and $^2A_1$ (and $^2A_2$) occur only once, and are hence represented by *straight lines,* having constantly a well-defined sub-shell configuration, $(t_2)^3$ for the $^4A_2$ and $(t_2)^2$ (e) for the three other (slope +1). All other curves are roots of secular equations of second (or higher) degree, and have asymptotic sub-shell configurations for very large Σ (shown as *stippled lines*)

octahedral symmetry have been elaborated in the classic work of Sugano et al.[18] and its relevance to LSC's has been discussed recently[19]. Figure 1 shows the ligand field of the Cr^{3+} energy levels in an octahedral coordination: on the left-hand side are the Russell-Saunders terms of free Cr^{3+}, and on the right-hand side the group theoretical designations for the symmetry sites in octahedral symmetry[19, 20]. The horizontal variable Δ/B, the ratio of the sub-shell energy difference Δ (= 10 Dq) to the Racah parameter B (separating Russell-Saunders terms such as 15 B between ^4P and ^4F of the d^3 (or the d^7) configuration). In this diagram, both the ground state $^4A_2$ and the excited state ^2E arise from the lower set (in octahedral symmetry) of d orbitals $(t_2)^3$ whereas $^4T_2$ belongs to $(t_2)^2(e)^1$.

In the spin-forbidden transition $^2E \rightarrow {}^4A_2$ there is no change in the geometry of the electronic orbitals and the equilibrium position of the ground and excited configuration is almost unchanged (see Fig. 2); the transition is sharp, the vibronic interaction weak and the situation of the R lines in ruby or alexandrite resembles that of rare earths. In contrast, the spin-allowed transition $^4T_2 \rightarrow {}^4A_2$ arises from the e set to the t_2 set. Since

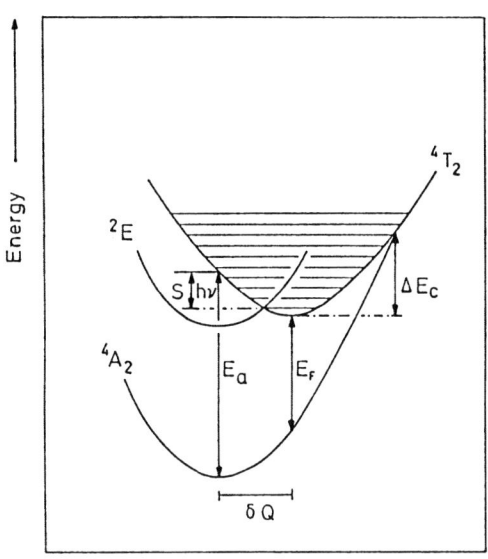

Fig. 2. Configuration coordinate diagram for the ground state $^4A_2$ and the excited $^4T_2$ and ^2E states for Cr^{3+} in intermediate fields

Configuration coordinate ⟶

the two e orbitals are pointed along the axes of the octahedron in the direction of the ligands, they participate in anti-bonding molecular orbitals, and the presence of one electron in these orbitals in the excited state causes a distortion and an increase in equilibrium internuclear distance between Cr^{3+} and the ligands. The extension of the electronic function results in its much stronger interaction with the vibrational modes. The situation is evident from the configurational diagram (see Fig. 2) and as described here is characteristic of the general behaviour of molecules and ions which exhibit strong vibronic coupling when the transitions are spin-allowed, and a weak coupling when the transitions are spin-forbidden.

Whether the energy of $^4T_2$ is higher than the energy of ^2E or vice versa in a given chromium compound depends on the strength of the ligand field acting on Cr^{3+}. This can be seen directly from Fig. 2 where $\triangle E$ represents the energy difference between the $^4T_2$ and ^2E states[21]. Experimentally it is measured from the position of the zero-phonon $^4T_2$ and ^2E bands. The values of both energies are equal and $\triangle E = 0$ for $\Delta/B \cong 20$ (intermediate field); high fields have Δ/B greater than 20 and $^4T_2$ above the ^2E level, and low fields have Δ/B less than 20 and $^4T_2$ below the ^2E level. Besides the variation in Δ with interatomic distance, the other atoms bound to the nearest-neighbour oxygen atoms may have some influence.

Glasses provide low field Cr^{3+} sites in which the zero-phonon $^4T_2$ level lies below the zero-phonon ^2E state. In these low field cases the room temperature Cr^{3+} emission from $^4T_2$ consists of a broad unstructured band centred in the near IR. From electrostatic considerations a more open structure arises from the larger distance between Cr^{3+} and its ligands and results in low fields. Quantum yields for luminescence are low in glass hosts; for example the highest quantum yield of Cr^{3+} found to date in silicate glass is 11%–17%. As already mentioned above the best efficiencies of 22% for $^4T_2 \rightarrow {}^4A_2$ are obtained for lithium lanthanum phosphate (LLP) glasses[11, 12].

6 Experimental Results

The resulting glass-ceramics obtained at various experimental conditions consist of a crystalline phase and a residual glassy phase. The nature of the crystalline phase corresponding to different heat treatment and precipitation conditions is determined by X-ray diffraction[5, 20–25]. This together with a detailed spectroscopic study of the steady state fluorescence, absorption, decay dynamics (by means of selective laser spectroscopy) as well as electron paramagnetic resonance reveals the detailed nature of the crystalline phases.

After determination of the crystalline phases of the glass-ceramics synthetic "real" crystals[2, 5, 22–25] are prepared by firing and annealing a mixture of stoichiometric amounts of the corresponding oxides. The crystals are characterized by steady state and time-resolved spectrosopy, X-rays and EPR spectrosopy. An analogy was found in the spectroscopic properties and high efficiency of fluorescence between Cr^{3+} in glass-ceramics and crystals[26].

6.a Types of Glass-Ceramics Studied

The glass-ceramics investigated were spinel-type[1, 24, 29], gahnite-type[1, 2, 23], a petalite-like phase[3], β-quartz[3], mullite[4, 22] and virgilite[5]. The corresponding minerals have the crystal type and (frequently rather idealized) chemical composition:

spinel: cubic $MgAl_2O_4$
gahnite (spinel-type): $ZnAl_2O_4$
petalite: monoclinic $LiAlSi_4O_{10}$
virgilite: hexagonal $Li_xAl_xSi_{3-x}O_6$
mullite: orthorhombic $Al_6Si_2O_{13}$
cordierite: orthorhombic $Mg_2Al_4Si_5O_{18}$

X-ray diffraction studies of glass-ceramics reveal similarities with the synthetic crystals[5, 22–26].

6.b Nucleation

Refractory oxides, such as titanium oxide, zirconium oxide[26] and iron(III) oxide[27] are frequently used. Chromium(III) can also serve as a nucleating agent as shown by Small Angle Neutron Scattering (SANS) and Small Angle X-ray Scattering (SAXS) in the case of cordierite glass, of the starting composition, 52 SiO_2, 34.7 Al_2O_3, 12.5 MgO and 0.8 Cr_2O_3[10], and also for formation of spinel-type glass-ceramics. These techniques give information about particle size. The nucleation of spinel occurs after diffusion of Cr^{3+} through the bulk of the glass. Heating of the glass at 850 °C for 2 hours results in spinel with particle size of 40 Å dispersed in the glassy phase. The particle size increases after additional heat treatment and may be as high as 280 Å[28].

6.c Steady State Spectroscopy

Steady state absorption allows determination of the ligand field strength acting on the Cr^{3+} ion. The absorption due to $^4A_2 \rightarrow {}^4T_2$ spin-allowed transition peaks around 650 to 560 nm[11]. As a result of stronger ligand field which is exerted on Cr^{3+} having shorter distances to its ligating oxygens in more rigid glass-ceramics than glasses the $^4A_2 \rightarrow {}^4T_2$ absoprtion peak is shifted to shorter wave-lengths, i.e. higher energies.

Table 1 presents ligand field parameters obtained from the absorption peaks of the $^4A_2 \rightarrow {}^4T_2$ transition of Cr^{3+} in glasses and glass-ceramics.

Comparison of the value in Table 1 shows that the ligand field of Cr^{3+} in the crystalline phase of the glass-ceramics is higher than in the glassy phases or in the glasses and its value is comparable to that of Cr^{3+} in ruby (and alexandrite). The ^2E level is 1000–3000 cm^{-1} below the minimum of the $^4T_2$ level. Such a situation permits thermal equilibration between two levels as has been found previously in many complexes and crystals.

6.d Steady State Emission

The emission of Cr^{3+} in glass-ceramics differs in shape and quantum efficiency from that of glasses. It is usually composed of a broad emission peaking around 750–900 nm arising from the $^4T_2 \rightarrow {}^4A_2$ spin-allowed transition which is usually excited at the absorption

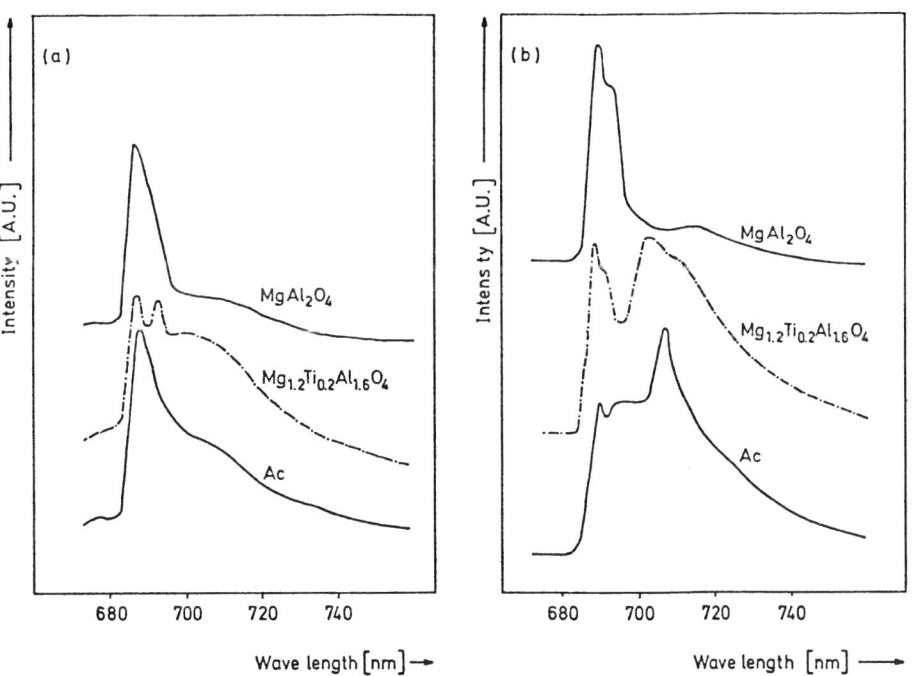

Fig. 3a, b. Emission spectra of different samples under 590 nm laser excitation. **a** room temperature; **b** 4.4 K

Table 1. Ligand field parameter Δ (in the unit 1000 cm^{-1}) for $Cr(III)O_6$ in glass-ceramics (g.c.); Solids and Complexes in Solution

Spinel-type g.c.	original glass	15.5	
-do.-	glassy micro-phase	15.95	
-do.-	micro-crystallites	17.25	
β-quartz g.c.	glassy micro-phase	16.0	
-do.-	micro-crystallites	16.8	
Petalite-like g.c.	original glass	15.6	
-do.-	glassy micro-phase	16.0	
-do.-	micro-crystallites	17.5	
Various silicate glasses			15.2–15.5
Lithium-lanthanum-phosphate glass			15.45
Alexandrite $Al_{2-x}Cr_xBeO_4$			17.35
Spinel $MgAl_{2-x}Cr_xO_4$			18.2
Spinel-type $MgAlCrO_4$			17.65
Ruby $Al_{2-x}Cr_xO_3$ (x < 0.04)			18.0
Greyish-pink $Al_{1.6}Cr_{0.4}O_3$			17.55
$Cr(C_2O_4)_3^{-3}$ [oxalate complex]			17.5
$Cr(OH_2)_6^{+3}$ in water			17.45

around 625 nm, and a sharp well-resolved emission around 680–700 nm due to $^2E \rightarrow {}^4A_2$ transitions from octahedrally distorted single Cr^{3+} sites and pairs. Figure 3 presents an example[24, 29] of emission spectra measured at room temperature and at 4 K for a spinel-type glass-ceramics, synthetic spinel crystal $MgAl_2O_4$ and mixed crystal of $Mg_{1.2}Ti_{0.2}Al_{1.6}O_4$.

The emission spectrum is composed of the $^2E \rightarrow {}^4A_2$ emission which is characteristic for Cr^{3+} in the crystals with a large field strength, and of the wide $^4T_2 \rightarrow {}^4A_2$ emission in the glass and low field crystalline sites. The $^2E \rightarrow {}^4A_2$ emission reveals a complex structure which is well resolved at 4 K[29]. In order to understand the origin of various bands we need to know the crystalline structure of spinels and sites into which Cr^{3+} can enter.

The usual "normal spinel" is $A^{(4)}B_2{}^{(6)}X_4$ where A is a divalent cation, B a trivalent cation and X a divalent anion. The symbol (4) designates tetrahedral coordination and (6) octahedral coordination. However spinels are known to undergo inversion, the formula of "inverse spinel" being $B^{(4)}[AB]^{(6)}X_4$. The degree of the inversion depends on the conditions of crystallization. Usually it is much larger in synthetic spinels than in natural spinels[30].

In the spinel structure the Cr^{3+} impurities occupy octahedral cation positions[31]. Actually, almost all Cr(III) complexes in solutions as well as the solid compounds are known to show coordination number N=6 with octahedral symmetry[6].

The spinels $MgAl_2O_4$ undergo inversion to some extent when heated between 750–900 °C (and higher temperatures) without annealing[30].

Such inversion must occur to a considerable degree in the glass-ceramics Ac as evident from the appearance of additional lines due to $^2E \rightarrow {}^4A_2$ emissions resulting from the modified Cr(III) energy levels in the low symmetry characterizing cases of inverted spinel sites. This behaviour in crystals was thoroughly studied by Mikenda et al.[30–33].

The intrinsic $^2E \rightarrow {}^4A_2$ luminescence (R-line) has a lower intensity, as typical for strictly octahedral symmetry, than the Cr(III) luminescence close to inverted sites[30]. This is not surprising, since the parity-forbidden transitions become partially allowed when the high symmetry is lowered. The R-line intensity increases when Cr(III) is excited at shorter wavelengts. A weak line at ~ 680 nm in the emission spectrum of T-51 glass-ceramics at 555 nm excitation is the R line[26]. This line is missing at 622 nm excitation. The next-most intensive line at 690–695 nm arises from Cr(III) in distorted octahedra CrO_6.

The so-called N lines[31-34, 93] either originate in such distorted sites, or are due to interactions between two or more Cr(III) at relatively short mutual distances, also implying anti-ferromagnetic coupling. The latter lines can usually be recognized by their increasing intensity as a function of higher Cr(III) concentration. In both cases[33, 93] the life-time of N line emission decreases systematically with their distance from R line emission on the regular octahedral sites, allowing differentiation by time-resolved spectroscopy (having the advantage that transitions to several components of $^4A_2$ from two or more Cr(III) have the same life-time from a given excited state). Also, differing clusters can be distinguished by the positions of the $^4T_2$ maxima[32] of excitation spectra for N line emission.

The intensive but almost hidden line at 706 nm, the N_4-line [31-34] is due to the exchange-coupled Cr^{3+} ions (emission of Cr^{3+} pairs, like the R line, changes its place only a little in the spectra of different spinels and many other crystals). The origin of this line from the Cr^{3+} pairs is deduced from lifetime measurements and time-resolved spectroscopy at low temperatures[29].

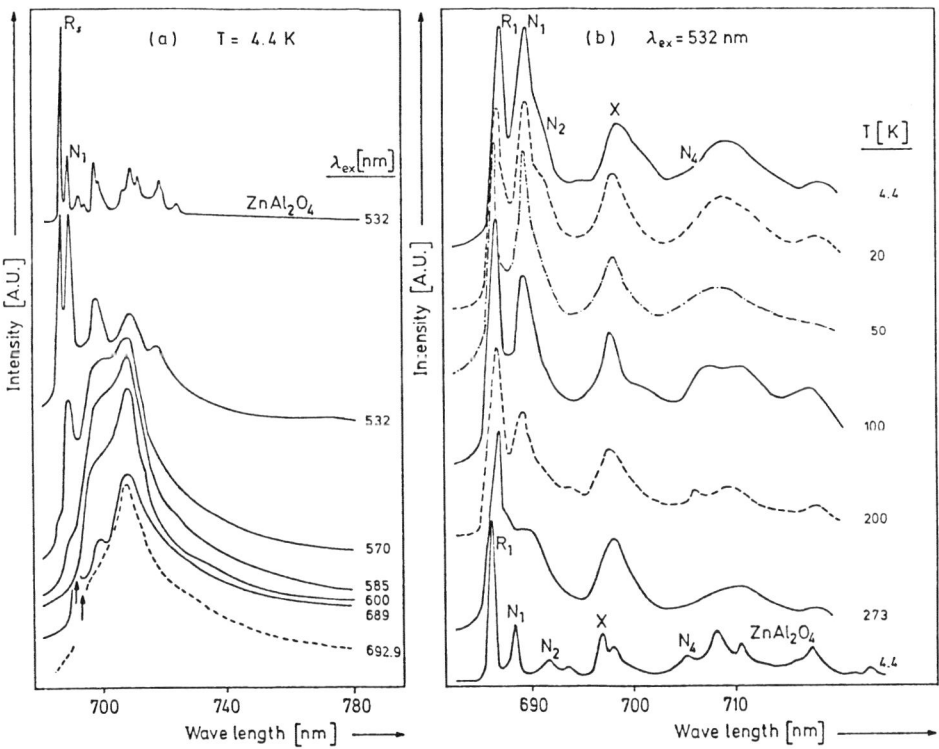

Fig. 4a, b. Comparison of the emission spectra of Cr^{3+} in $ZnAl_2O_4$ at $\lambda_{ex} = 532$ nm and temperature 4.4 K with Cr^{3+} in gahnite glass-ceramics. **a** at various excitation wavelengths and temperature 4.4 K; **b** at $\lambda_{ex} = 532$ nm and various temperatures

The gahnite glass-ceramics samples are prepared from a glass of starting composition 70.2 SiO_2, 15.0 Al_2O_3, 4.4 ZnO, 7.1 Li_2O, 1.5 TiO_2, 1.5 ZrO_2, 0.3 As_2O_3 mole% as described in reference 1. The specific glasses are doped by Cr^{3+} at a concentration of 1.1×10^{19} ions/cm^3.

The powdered crystalline samples of Cr^{3+} doped $ZnAl_2O_4$ gahnite were prepared for comparison from stoichiometric ratios of $Zn(NO_3)_2$, $Cr(NO_3)_3$ and $Al(NO_3)_3.9 H_2O$ precipitated by aqueous ammonia, and the mixed hydroxides calcined. The lattice constant **a** of the crystalline $ZnAl_2O_4$ was found by X-rays to be 8.085 Å.

Figures 4a and 4b present a comparison of the emission spectrum of Cr(III) in gahnite-like glass-ceramics at various excitation wavelengths at 4.4 K. We observe the R_1-line of Cr(III) in gahnite-like glass-ceramics at various excitation wavelengths at 4.4 K. We observe the R_1-line of Cr(III) at 686.2 nm both in $ZnAl_2O_4$ crystal and in glass-ceramics in contrast with $MgAl_2O_4$ synthetic crystal and glass-ceramics where no such line appears[24]. The appearance of these lines is an evidence of Cr^{3+} in almost octahedral symmetry with no adjacent inverted site by analogy with the work of Mikenda et al.[31-33]. The N_1 (688.6 nm) line seen clearly both in the crystal and in glass-ceramics is actually ascribed to distorted Cr^{3+} sites. N_2 (692.3 nm) has much lower intensity than N_1 and is seen clearly in the crystal but it obscured in the glass-ceramics. X (~ 697.0 nm) is of comparable intensity with the N_1 line in the crystal and lower than the N_1 line in glass-ceramics. It has been shown by Mikenda to decrease with the concentration of Cr^{3+} ions[31-33]. Therefore it has to be ascribed to singly perturbed ions. N_4 group of lines (705.0 to 715.0 nm) can be ascribed to distorted Cr^{3+} pairs. In the crystal three distinct lines can be seen and one broad band in the glass-ceramics as evident from Fig. 4b. The intensity of these bands increases at lower temperature due enhanced association of Cr(III) into pairs. The existence of such pairs was also detected by EPR measurements.

The excitation spectra of gahnite glass-ceramics due to $^4A_2 \rightarrow {}^4T_2$ broad-band absorption at 4.4 K are given in Fig. 2 of Ref. 2. The excitation peaks shift to lower energies in the order of selected emission wavelengths (given in parentheses):
R_1 line (686.2 nm)
N_1 line (688.6 nm)
X lines (701.2 nm)
$^4T_2 \rightarrow {}^4A_2$ (730 nm; maxima in excitation spectrum at 590 and 620 nm) same broad band (790 nm; maximum in excitation at 630 nm) which is consistent with the fact that the more regular Cr(III) sites are showing higher, and the pairs somewhat lower, field-strengths.

Figure 5 shows several time-resolved emission spectra at 4.4 K under 570 nm excitation. The R-line has the longest time-constant, the pairs exhibit the shortest time-constant as expected, while the perturbed ions are the intermediate case. The slopes of the semilog decay curves for the crystal are exponential in most cases, with slight deviation of the exponentiality at very short times. In the glass-ceramics the deviation from exponentiality occurs at much longer time intervals. The decay time constants may be obtained from the exponential portions of the curves. The extremely long lifetime of 33 ms in the crystal at 4.4 K is indicative of the very high symmetry in which Cr^{3+} is situated. It also results from energy transfer from the R-line to pairs and to perturbed ions. The fact that the long time portion of the decay curves are almost parallel indicates that the survival probability of excited state population of the regular octahedral sites controls the kinetics of the decay.

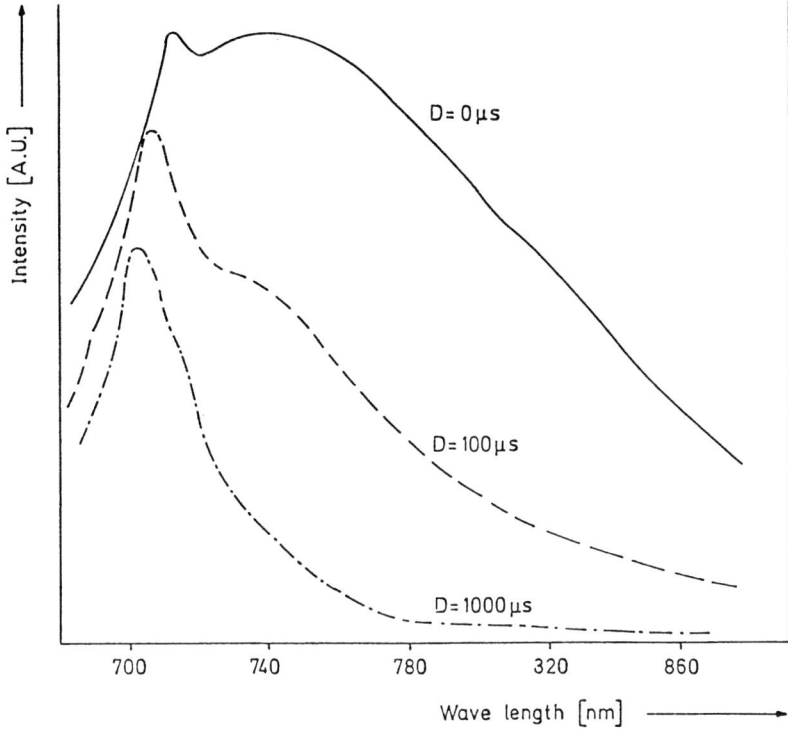

Fig. 5. Time-resolved spectra of spinel-type glass-ceramics A_c excited at 680 nm and at room temperature. The gate width is 100 µs, D is the time delay

The variety of emission lines in the gahnite-like glass-ceramics and the possibility to ascribe these lines to various Cr^{3+} centers can be a useful tool in following the mechanisms of the nucleation of glass-ceramics doped by Cr(III). The high quantum efficiency of Cr(III) emission in glass-ceramics and the proximity of $^4T_2$ and ^2E levels in the distorted sites may have importance in designing tunable lasers.

7 Time-Resolved Spectroscopy

The time-resolved spectroscopy of Cr^{3+} doped spinel and petalite-like phase types glass-ceramics[5, 23, 24, 29] enables a more accurate distinction of the ^2E and $^4T_2$ energy levels of Cr(III) and of the equilibrium between the population of two levels[29], see Fig. 5, which presents the time-resolved spectra of spinel type glass-ceramics excited at 680 nm at room temperature[1, 24]. Glass-ceramics derived from boro-silicate glasses containing Cr(III) were also recently studied[135] by laser spectroscopy and fluorescence-line narrowing.

Immediately after excitation we see a broad band emission peaking at around 740 nm due to the $^4T_2 \rightarrow {}^4A_2$ transition. After longer times this emission decreases and a long-lived emission from the metastable state $^2E \rightarrow {}^4A_2$ peaking around 700 nm is observed.

The ^2E emission can be further resolved when the emission spectra are studied at liquid helium temperature, when the R line peaking at 685 nm is assigned to single chromium ions and the one peaking at 705 nm to antiferromagnetically coupled pairs of Cr^{3+} [2].

By comparing the emission spectra of Cr^{3+} in glass-ceramics with that of Cr^{3+} in single crystals one finds that the tendency to form pairs is greater in the glass-ceramics. In mixed crystals of spinel the appearance of Cr^{3+} pairs is again significant.

8 Electron Paramagnetic Resonance

Cr^{3+} has three unpaired d electrons in its ground state and can be easily detected by the signal in the EPR spectrum.

The EPR spectra of Cr^{3+} in glasses have been studied extensively by Landry et al.[35] who performed a thorough analysis of EPR spectra of Cr(III) in glass and attribute the diffusive absorption at 1000–1500 Gauss to distorted octahedrally coordinated single Cr^{3+} ions [10, 18, 28, 35, 36].

In oxide glasses Cr(III) gives rise to a very characteristic resonance spectrum for which the maximum of the derivative corresponds to an effective Landé factor g of 5 to 6, i.e. a magnetic field H between 1 kG and 1.3 kG at 9.05 GHz [35, 36]. Such a signal arises mainly from the transition $-3/2 \rightarrow +3/2$ which is allowed due to the low symmetry of the Cr^{3+} sites in glass. The signal of Cr^{3+} in the glass is therefore easily distinguished from that of Cr^{3+} in crystalline phases because of its position and characteristic shape.

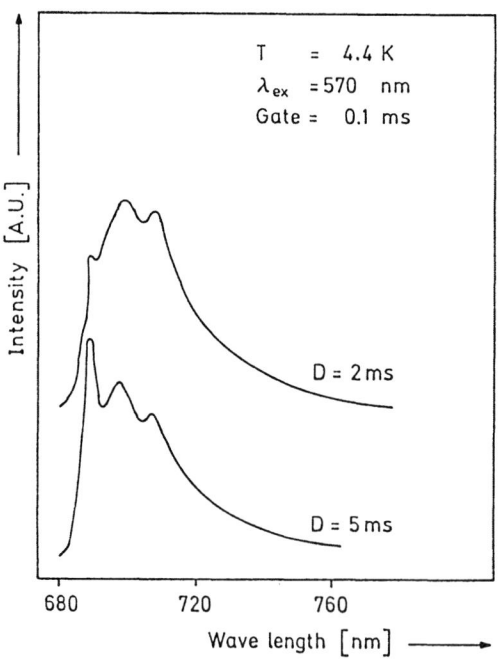

Fig. 6. Time-resolved spectra of Cr^{3+} in gahnite glass-ceramics at $\lambda_{ex} = 570$ nm and temperature 4.4 K

The EPR spectra in the crystalline phase are explained by analogy with the work of Durville et al.[28] who attribute the sharp absorption at 1580 Gauss to Cr^{3+} in high symmetry sites of the crystalline phases. The appearance of 1460 G narrow peaks in all samples is due to the ferric impurity[37] and at 3300 Gauss, to pairs.

Figure 6 presents the EPR spectra of Cr^{3+} in gahnite-type glass-ceramics under different heat treatments. This type of glass-ceramics was prepared from glass of original composition (mole %) 73.6 SiO_2, 11.8 Al_2O_3, 4.2 Li_2O, 7.0 ZnO, 1.6 TiO_2, 1.5 ZrO_2, 0.3 As_2O_3, 0.024 Cr_2O_3 (g/c 1_2) by heating at 750 °C for 10 h and then at 860 °C for 2 h and at 890 °C for an additional 2 h. Sample g/c 1_3 was heated at 750° for 10 h and then at 900 °C for 1.5 h and finally at 910 °C for 2 h. The crystalline phases determined by X-ray diffraction consisted of a solid solution of ZrO_2; β-quartz and gahnite.

9 Recent Progress in Theory

The absorption spectra of compounds containing a partly filled 3d, 4d or 5d shell were to a large extent[18, 20, 38–40] rationalized by "ligand field" theory applied to aqua ions $M(OH_2)_6^{+z}$, ammonia complexes $M(NH_3)_n(OH_2)_{6-n}^{+z}$, complexes of multidentate (synthetic or biological) aminopolycarboxylate anions such as ethylenediaminetetra-acetate M(edta) $(OH_2)_x^{+z-4}$ and glycinate $M(NH_2CO_2)_n(OH_2)_{6-2n}^{+z-n}$ in solution, for which a rich material of formation constants were available, even before 1960. This does not prevent that fully convincing evidence for the stoichiometry and local symmetry of aqua ions, and many other complexes, in solution, could only be obtained by comparison with the reflection and/or transmission spectra of salts with known crystal structure, such as M(II) Tutton salts (schoenites) $A_2[M(OH_2)_6]$ $(SO_4)_2$ and M(III) alums $A[M(OH_2)_6]$ $(SO_4)_2$, 6 H_2O where A are alkali-metal or ammonium ions. Partly related to the operation in 1960 of the first laser [ruby $Cr_xAl_{2-x}O_3$] much work was initiated on absorption (and in fortunate cases luminescence spectra) of d-group ions incorporated (usually at low concentration) in colourless crystals of known structure[41–43]. This corroborated an interest in explaining the bright colours of many minerals, though a major difficulty here can be grey to black coloration due, for instance, to the simultaneous presence of iron(II) and iron(III) in (otherwise colourless) silicate minerals, producing strong electron transfer bands[20]. Analogous to Prussian Blue $AFe^{II}(CN)_6Fe^{III}$ such colours can also be intensely blue, like in sapphire $Fe_x^{II}Ti_x^{IV}Al_{2-2x}O_3$ though simultaneous presence of titanium(III) and Ti(IV) usually provide brown-purplish colours of glass-ceramics[1] and of Al_2O_3[44]. Glasses are in much the same situation as solutions; though EXAFS can provide valuable information about the distances (with precision around 1%) from a given element (even at moderate concentration) to neighbour nuclei of other elements, the optical spectra of 3d group ions in glasses have brought relatively undramatic conclusions about the local symmetry; both Cr(III) and Ni(II), also in fluoride glasses[45], remain close to regular octahedral, and the major variation, like in crystalline mixed oxides[6, 20, 43] seems to be the one-electron energy difference Δ [between the two anti-bonding d-orbitals having angular functions proportional to $(x^2 - y^2)$ and to $(3z^2 - r^2)$, and the three roughly non-bonding orbitals (xy), (xz) or (xy)] decreasing some 5%[46, 47] for each percent increase of the chromium-oxygen internuclear distance R. The clear-cut (but perhaps not unique) rationalization for octahedral symmetry is that d^3 systems such as Cr(III) have no anti-bonding electron, compared to the average value 4 q/10 = 1.2 electron in a d^q system with spherically symmetric groundstate [like Mn(II) and Fe(III) for q = 5 and S = 5/2]. Also d^8 systems such as octahedral nickel(II) have two anti-bonding electrons rather than the average value 3.2, again avoiding 1.2 anti-bonding electron. This situation is entirely different in the 4f group, where the "ligand field" effects are 10 to 50 times weaker[48] than in the d-groups, and where the preference for a given coordination number N and a specific stereochemistry is far less pronounced, both in glasses[49] and in crystals[50, 51].

9.a Intermediate Coupling in the 11 Kramers Doublets Formed by $^4T_2$, ^2E and $^2T_1$

The ruby $Cr_xAl_{2-x}O_3$ (for small x) is an almost ideal material[41, 42] for detecting the doublets, ^2E (14418 and 14447 cm^{-1}) and $^2T_1$ at slightly higher energy, well below the first broad, stronger transition to $^4T_2$ (having the maximum at 18100 cm^{-1}), and the last sharp transition to $^2T_2$ at 21000 cm^{-1} situated between the latter quartet, and the next one ($^4T_1$) at 24400 cm^{-1}. In solution, ^2E can be observed[52] as a sharp, weak band at 14350 cm^{-1} (697 nm) in the oxalate complex $Cr(O_2C_2O_2)_3^{-3}$ and in the aqua ion as a shoulder at 14950 cm^{-1} (669 nm). Around 1960, it was felt that the spin-forbidden transition to ^2E would become very difficult to find, when closer to $^4T_2$ (not so much because it is weak in an absolute sense, as because it broadens and at the same time is superposed the far stronger $^4T_2$ band). Nevertheless, the urea complex $Cr(OC(NH_2)_2)_6^{+3}$ shows a highly complicated contour of the transition to $^4T_2$[53, 54] and the most clear-cut example, as far the crystal structure goes, is the cubic elpasolite-type K_2NaCrF_6 where two minor, sharp peaks indicate ^2E and $^2T_1$ at lower wave-numbers than 16200 cm^{-1} of the maximum of $^4T_2$, and a shoulder $^2T_2$ at 22000 cm^{-1} [55]. However, it is also clear that Cr(III) in such cases behaves rather similar to nickel(II) showing crossing of ^1E with $^3T_1$ at lower, and with $^3T_2$ at higher Δ values[56, 57]. Griffith[58] pointed out that in such situations, the double-group quantum numbers Γ_J replace the characterization of energy levels by the (upper left superscript) multiplicity $(2S + 1)$ and the symmetry types (such as A_1, A_2, E, T_1 and T_2 in the point-group O_h) exactly like monatomic entities conserve J when deviations from Russell-Saunders coupling no longer allow defined S and L values[59].

In gaseous Cr^{+3} the Landé parameter ζ_{3d} is close to 270 cm^{-1} as can be seen from the width 956 cm^{-1} of the lowest term ^4F (having the first-order width[59] $(3 + 1/2)\zeta_{nl}$). This Landé parameter is nearly twice the ζ_{2p} = 150 cm^{-1} in the oxygen atom, and almost half as large as ζ_{3d} = 630 cm^{-1} in gaseous Ni^{+2}. However, this is still much less than ζ_{5d} in $5d^3$ rhenium(IV) hexahalide complexes[60] and the isoelectronic gaseous molecule IrF_6 all quite close to 3000 cm^{-1}. These 5d group examples show that chemical bonding decreases the effects of spin-orbit coupling relative to the gaseous ion, much in the same way as the nephelauxetic effect[20, 61] is the decrease of the S.C.S. (Slater-Condon-Short-ley) parameters of interelectronic repulsion[59], the values for gaseous Cr^{+3} being multi-plied by a factor between 0.9 for the least covalent fluorides[45] to 0.4 for the most covalent Cr(III) compounds. Frequently, the increase of S.C.S. parameters with ionic charge is cancelled by the more extensive covalent bonding in the higher (isoelectronic) oxidation state, as known from many instances of chromium(III) and manganese(IV) in similar environments[20] or the $3d^5$ manganese(II) and iron(III). On the whole, the major effect of deviations from Russell-Saunders coupling in octahedral d-group complexes is intensification of spin-forbidden transitions, as seen in $4d^6$ rhodium(III) and somewhat stronger absorption bands[39, 62] in $5d^6$ iridium(III) which can be compared with the determinants of intermediate coupling in O_h[63]. In cases, where the ligands have much larger ζ_{np} (such as 2460 cm^{-1} for bromide and 5000 cm^{-1} for iodide) the moderate delocalization of d-like orbitals in the vicinity of the halide nuclei may intensify spin-forbidden absorption bands of $Cr(NH_3)_5 I^{+2}$ and $Co(NH_3)_5 I^{+2}$ as can also be discussed in the related case[64] of spin-orbit coupling influencing electron transfer spectra of octahedral osmium(IV) complexes of one or more types of halide ligands.

Table A 34, p. 419 of the book by Griffith[58] gives the non-diagonal elements (as multiples of ζ_{nd}) of intermediate coupling within the manifold consisting of the ground-state $^4A_2$, the excited $^4T_2$ and $^4T_1$ and the three lowest doublets ^2E, $^2T_1$ and $^2T_2$. The Bethe double-group quantum numbers (in the point-group O_h) Γ_6, Γ_7 and Γ_8 are called E' (characterizing $J = 1/2$ originating in spherical symmetry), E'' and (the two degenerate Kramers doublets) U (characterizing $J = 3/2$) by Griffith. The other J-value 5/2 feasible for a single d electron corresponds to both E'' and U. For our purposes, the most important area is the near-crossing of $^4T_2$ with ^2E and (the closely adjacent) $^2T_1$. The 22 $= 12 + 4 + 6$ states are distributed on 11 Kramers doublets forming seven Γ_J levels, two E', one E'' and four U. According to second-order perturbation theory, the intensification of spin-forbidden transitions is proportional to the square of the non-diagonal element, and inversely proportional to the distance (if many times ζ_{3d}) squared, between the doublet and the closest quartet component. It is here important that the squared non-diagonal element in the unit $(\zeta_{3d})^2$ is only above 0.25 in two cases, 4/3 between U(^2E) and U($^4T_2$); and 1/2 between E' ($^2T_1$) and E'($^4T_2$) playing a rôle for the intensity of the sharp band in the red. The minimum distance between two interacting components is twice the non-diagonal element, and hence only 600 cm^{-1} for the two U components in the case of coinciding diagonal energies of ^2E and $^4T_2$ (assuming the Landé parameter of gaseous Cr^{+3}). Actually, salts of CrF_6^{-3} and $CrF_5(OH_2)^{-2}$ [53, 65, 66], chromium(III) in strong solutions of sulphuric acid[53, 67] and a lot of differing glasses[21, 35, 67–69] show a structured absorption band in the red, looking like the superposition of several narrow peaks on a main band shaped like the usual Gaussian error-curve[38]. When one peak is at distinctly lower wave-number, there is no doubt that it is the lowest U level[58] consisting mainly of ^2E having the squared non-diagonal element 1.33 $(\zeta_{3d})^2$ with $^4T_2$ and hence some 0.5 to 8 percent quartet character according to the distance 4500 to 1100 cm^{-1} to the $^4T_2$ diagonal elements of energy. Though these ^2E and $^2T_1$ levels considered as $(t_2)^3$ have exactly the same d^3 interelectronic repulsion relative to $^4A_2$ in the S.C.S. treatment (in close analogy[58] to the ten states of ^2D in the configuration p^3), $9 B + 3 C \sim 21 B$ in terms of Racah parameters, it was pointed out by Walter Schneider[20] that the second-order perturbation from the higher sub-shell configurations $(t_2)^2e^1$ and t_2e^2 [18] is markedly stronger (though not exactly as different as suggested by the sum of the squared non-diagonal elements $90 B^2$, assuming $C = 4 B$) for ^2E than ($24 B^2$) for $^2T_1$. Hence, ^2E is slightly below $^2T_1$ on Fig. 1 though they approach asymptotically the same height over the groundstate. In cases like ruby, a lot of weak bands at energies above ^2E may either be vibronic co-excitations (like in ReX_6^{-2} [60] and IrF_6) of ^2E and/or $^2T_1$, and one band may represent the $^2T_1$ origin. Adamson[70, 71] substituted deuterium in $Cr(ND_3)_6^{+3}$ and could discern the vibronic components, and most probably detected the $^2T_1$ zero-zero line at 15 400 cm^{-1} to be compared with ^2E at 15 180 cm^{-1}. We return below (9b) to the evidence that a few

Cr(III) complexes may have $^2T_1$ below ^2E. The non-diagonal elements of spin-orbit coupling vanish[58] between the components of the adjacent ^2E and $^2T_1$.

Schäffer pointed out[66] that a superposition of broad Gaussian error-curves with the *same* one-sided half-width δ do not have a sum showing conspicuous shoulders, as long the mutual distance is well below one-half δ (and as a matter of fact, cases like spin-allowed transitions to T_1 and T_2 levels of Cr(III) and Ni(II) hardly differ from Gaussian shape, except perhaps as a weak asymmetry in δ toward lower compared to higher energy). On the other hand, even rather weak bands with distinctly lower δ are quite perceptible, and readily produce relative maxima and minima on the sum curve. Many authors tend to describe such phenomena as Fano anti-resonances (known from sharp atomic-like lines superposed a continuous slope) but we doubt that the intensities involved here are sufficient to induce such behaviour. Though glasses obviously have non-equivalent sites[49] the absorption spectra show a remarkably continuous development of three sharp peaks traversing the broad $^4T_2$ band, as if each case was fully determined by a definite value of $\Sigma = \Delta/B$ in the interval 17 to 25. This situation is closely similar to the crossing of ^1E and $^3T_1$ in numerous octahedral nickel(II) complexes[57] and would raise hopes that a comparable atlas of chromium(III) band shapes could be established[65-69] as indicator of Σ, the ratio Schäffer calls the "ligand field strength".

9.b Mono-Axial Distortions and Specific Results for Orthoaxial Chromophores

In the early model[58] ascribing energy differences between the five d-orbitals to the non-spherical part $V(x, y, z) - V(r)$ of the Madelung potential [the d-group nucleus at $(0, 0, 0)$] a strong source of discrepancies with observed spectra was the prediction that small distortions of a regular octahedron having the sub-shell energy difference Δ proportional to $\langle r^4 \rangle / R^5$ (the average value of r^4 for the d orbitals, divided by the internuclear distance R in the fifth power) should result in dramatic variations of band positions, since the major part of non-cubic perturbations are proportional to $\langle r^2 \rangle / R^3$. A consequence of the old model remains valid[40, 72] that only the *holohedrized symmetry* $[V(x, y, z) + V(-x, -y, -z)]/2$ influences the energy levels (we do not here speak about band intensities and luminescent transition probabilities) so the holohedrized point-group (which cannot be lower than C_i) determines the band splittings. With exception of C_i and orthorhombic D_{2h}, all non-cubic holohedric point-groups are *mono-axial*, the C_n axis (for $n = 2, 3, 4, 5\ldots$) combined with additional symmetry elements to form C_{nh} and D_{nh} (all in the case of n even), S_{2n} and D_{nd} (in the case of n odd, at least 3). If all the nuclei (considered as geometrical points) are situated on one straight line, the simultaneous presence of all C_n axes produces the linear holohedrized symmetry $D_{\infty h}$.

Claus Schäffer pointed out to us that the word "uniaxial" means something slightly different from the neologism "mono-axial" [Greeks would say the less transparent "monaxial"] in crystallography. Contrary to cubic and other isotropic (vitreous) solids, uniaxial crystals have *one* axis in the only direction where no double refraction occurs (in all other directions due to differing refractive index for light polarized in two perpendicular directions). However, in sufficiently low symmetry, crystals can be biaxial, having two axes without double refraction, as known from the anisometric class constituted by crystals of the orthorhombic, monoclinic and triclinic systems. Among the holohedrized non-cubic point-groups, all mono-axial cases also correspond to uniaxial behaviour of light refraction, with the one exception of C_{2h}. Actually, uniaxial systems have one (and not more) C_n axis (n = 3, 4, . . .) though a series of C_x axes may coincide in space, when x are all the divisors of the highest (x = n).

In agreement with the ideas of Alfred Werner, nearly all spectroscopic evidence[39] is compatible with at least the six closest neighbour nuclei being situated on the axes of a suitably chosen Cartesian system in octahedral complexes of $3d^3$ chromium(III), $3d^8$ nickel(II)[56, 57], $4d^6$ rhodium(III) and $5d^6$ iridium(III)[62]. Such chromophores MX_6 (and for that matter also quadratic MX_4) are called[73] orthoaxial (this is a property of a definite chromophore, but not a general property of a given point-group). The holohedrized symmetry of an orthoaxial chromophore (not having all nuclei on one straight line) can only be O_h, D_{4h} or D_{2h} (colloquially called octahedral, tetragonal and orthorhombic). If two ligands A and B differ (or is the same ligand at different distance R), the holohe-drized symmetry of MA_5B and both cis- and trans-MA_4B_2 is D_{4h}, of mer-MA_3B_3 (like most of the orthoaxial chromophores with more than two kinds of ligands; six different ligands produce the point-group C_1 before holohedrization) is D_{2h} and of fac-MA_3B_3 is O_h [the resulting d^q energy levels are identical with those of regular octahedralMX_6 of a hypothetical ligand X = (A + B)/2]. If at least two Cartesian axes are occupied by nuclei, orthoaxial chromophores realizing mono-axial point-groups by holohedrization can only have the tetragonal holohedrized symmetry D_{4h}.

The angular overlap model (A.O.M.)[40, 46–48, 73–76] rationalizes most of the observed effects of non-cubic distortions of an approximately octahedral chromophore. This is particularly true for tetragonal orthoaxial chromium(III) complexes[77]. However, the original assumption (cf.[48, 50]) of exclusively σ-anti-bonding A.O.M. effects must be extended to the possibility of moderate π-anti-bonding effects (disregarding for a moment π-back-bonding to empty ligand orbitals, which seems only to be significant in certain non-aliphatic organic ligands, including CN^- and CO). Hence, Δ of regular octahedral MX_6 becomes the difference between the σ-anti-bonding effect on $(x^2 - y^2)$ and $(3z^2 - r^2)$ angular functions and the π-anti-bonding effect on (xy), (xz) and (yz) orbitals. If we orient the Cartesian axes x and y through the A nuclei in trans-MA_4B_2, the σ-anti-bonding effect of the B ligands acts only on the cylindrically symmetric $(3z^2 - r^2)$ [symmetry type a_{1g} in D_{4h}] and the π-anti-bonding effect of B on (xz) and (yz) [these two orbitals have the symmetry type e_g in D_{4h}]. Taken together, the four A ligands make $(x^2 - y^2)$ [symmetry type b_{1g}] as σ-anti-bonding as it would be in MA_6, but $(3z^2 - r^2)$ only a-third as σ-anti-bonding (the remaining two-thirds being due to the ligands on the z-axis). The four A ligands, if they have any π-anti-bonding effect, make (xz) and (yz) each half as π-anti-bonding as (xy) [symmetry type b_{2g} in D_{4h}]. If π-anti-bonding predomin-antly is due to B ligands, the order of one-electron energies is $(b_{2g}) < (e_g) \ll (a_{1g}) \sim (b_{1g})$. The relative order of the two last orbitals depends on whether A or B is the stronger σ-anti-bonding. A comparative study[77] of several trans-Cr $(NH_3)_4X_2^+$ provided numerical values of these effects. F^- and OH^- are 30% as π-anti-bonding as they are σ-anti-bonding. OH_2, Cl^- and Br^- are decreasingly π-anti-bonding. If it is assumed that NH_3 is exclusively σ-anti-bonding, it turns out (as expected) that primary aliphatic amines also are, but a weak residual effect of π-back-bonding can be perceived in pyridine. The conventional spectrochemical series of Δ variation is $Br^- < Cl^- < F^- < OH^- < OH_2 < NH_3$ but the series of σ-anti-bonding alone is $Br^- < Cl^- < OH_2 < NH_3 < F^- < OH^-$. Unfortunately, no monomeric chromium(III) complexes are known of oxo ligands (because of their high proton affinity) but for our purposes, it is likely that typical oxo ligands are, at the same time, stronger σ- and π-anti-bonding than fluoro ligands. The Δ values measured of solid oxides (cf. Table 1) are either above the aqua ion (spinel-type $MgAl_{2-x}Cr_xO_4$; alexandrite $Cr_xAl_{2-x}BeO_4$ discussed below, 9d; corundum-type $Cr_xAl_{2-x}O_3$) or below[43] (corundum-type Cr_2O_3 and $Cr_xGa_{2-x}O_3$; perovskite-type $LaCr_xAl_{1-x}O_3$ and $YCr_xAl_{1-x}O_3$) but surprisingly slightly varying (like the practically identical Δ for $Cr(OH_2)_6^{+3}$ and the tris-oxalato complex).

The one-electron energy differences[77] in $CrA_4X_2^+$ are not directly accessible from the observed absorption maxima, but need closer analysis. This is also true for Δ of MX_6 containing from two to eight d-like electrons[20, 38, 58] with a few fortunate exceptions, such as the first spin-allowed transition in d^3 and d^8. If $trans$-$CrA_4X_2^+$ has distinctly π-anti-bonding X^- ligands, the $^4T_2$ level splits in ^4E_g at lower and $^4B_{2g}$ at higher energy (for convenience rather than for group-theoretical perfection, we keep the indicator "g" of even parity in D_{4h} but not in O_h to distinguish the two categories of quantum numbers) whereas the subsequent $^4T_1$ in most cases rather shows $^4A_{2g}$ and ^4E_g roughly at the same energy. (For various unexpected reasons[73] the same behaviour is shown by d^6 cobalt(III) and rhodium(III), the first spin-allowed transition to $^1T_1$ separating into ^1E_g at lower and $^1A_{2g}$ at higher energy; and $^1B_{2g}$ and ^1E_g usually remaining rather close to the position $^1T_2$ would have had with a hypothetical $(2A + X)/3$ ligand. Among the d^3 levels in tetragonal symmetry, $^4B_{2g}$ belongs exactly to the configuration $(e_g)^2 (b_{1g})$ and $^4A_{2g}$ to $(e_g)^2 (a_{1g})$ whereas the two ^4E_g are mixtures of $(b_{2g}) (e_g) (a_{1g})$ and $(b_{2g}) (e_g) (b_{1g})$. Much in the same way as the J-levels of a partly filled 4f shell undergo a tiny "ligand field" influence (hardly mixing the J-levels in most cases)[48, 50], the mono-axial (in the example tetragonal) deviation from cubic symmetry is a minor modification, and it is more judicious to "pre-form" the linear combination of the two M.O. configurations adapting the two ^4E_g levels to be appropriate members of the cubic levels $^4T_2$ and $^4T_1$. This is the motivation behind the recent concept[78] of "rediagonalization". Our discussion of $^4T_1$ assumes tacitly that the second $^4T_1$ at much higher energy almost exactly belongs to the cubic sub-shell configuration $t_2(e)^2$ and hardly has any importance for the visible spectrum. Because of the much lower ratio $\Sigma = \Delta/B$ in nickel(II) complexes, the analogous two $^3T_1$ are mixed with comparable squared amplitudes of $(t_2)^5(e)^3$ and of $(t_2)^4(e)^4$ and hence, it is convenient to rediagonalize for this purpose[78].

The *theorem of Tanabe and Kamimura* (p. 398, Ref. 79; p. 333, Ref. 40) is that a half-filled set of degenerate orbitals (i.e. identical one-electron energy) has energy levels that not split to the first order (but at the most, parabolically) as a function of an added perturbation. This is true for d^5 not only for spin-orbit coupling separating very slightly the J-levels of the quartet terms, but also for a cubic "ligand field". The Tanabe-Sugano diagrams[18] are identical for ($-\Delta$ and d^{10-q}) compared with Δ and d^q (as early pointed out for the lowest term by Van Vleck) and hence, d^5 is self-conjugate. The groundstates $^4A_2$ of d^3 and $^3A_2$ of d^8 are Tanabe-Kamimura stable with respect to mono-axial distortions; we have discussed above the tetragonal holohedrized symmetry providing $(e_g)^2 b_{2g}$ and $(e_g)^4 (b_{2g})^2 (a_{1g}) (b_{1g})$ having the same wave-functions as the cubic groundstates with sub-shell configurations $(t_2)^3$ and $(t_2)^6 (e)^2$. However, the Tanabe-Kamimura theorem is not restricted to the groundstate with maximum spin quantum number S. In the tetragonal M.O. configuration $(e_{2g})^2 (b_{2g})^1$ the twelve states form $^4B_{1g}$ (the cubic groundstate $^4A_2$), $^2A_{1g}$, $^2A_{2g}$, $^2B_{1g}$ and $^2B_{2g}$. Each of these levels correspond to a definite excited level in cubic symmetry, $^2A_{1g}$ and $^2B_{1g}$ constitute ^2E; $^2A_{2g}$ is a part of $^2T_1$; and $^2B_{2g}$ is a part (one of the three Kramers doublets) of $^2T_2$. The four states of each of the M.O. configurations $(e_g)^3$ and $(e_g)(b_{2g})^2$ form a level ^2E_g. Hence, close to cubic symmetry, the cubic levels $^2T_1$ and $^2T_2$ have each such a ^2E_g with equal squared amplitude of both tetragonal M.O. configurations, and they do not deviate (in first-order) from their partners $^2A_{2g}$ and $^2B_{2g}$, respectively, all eight Kramers doublets of ^2E, $^2T_1$ and $^2T_2$ having one-third (b_{2g}) and two-thirds (e_g) electronic density, exactly like the groundstate. The deviations of the two ^2E_g from the horizontal lines on the figure (p. 5013 of Ref. 80) are hence parabolic to the left of the figure, and not linear slopes. Broad emission from the ^2E_g component originating in $^2T_1$ occurs[81, 82] at $14\,140$ cm^{-1} in trans-$Cr(NH_3)_4F_2^+$ and between $12\,940$ and $12\,650$ cm^{-1} in analogous complexes of two bidentate or one cyclic quadridentate amine. The analogous cis-cases show narrow-band ^2E emission somewhere in the interval $15\,080$ to $14\,580$ cm^{-1}. According to private communication from Claus Schäffer, spin-orbit coupling plays a very minor rôle during the crossing of ^2E with the $^2T_1$ low-energy component. These results seem to show that the $^2T_1$ is slightly above ^2E in the absence of tetragonal distortion. Rather scanty evidence is available for the order of magnitude of $^2T_1 - {}^2E$ distances being 700 cm^{-1} in ruby[42] and $Cr_xGa_{2-x}O_3$[83], in tris(dipyridyl) and tris(phenanthroline) complexes[84] and a few other cases all sharing the problem of conceivable mis-identification of vibronic components as $^2T_1$. The "reineckeate" anion trans-$Cr(NH_3)_2(NCS)_4^-$ has four sharp absorption bands at $13\,400$, $13\,770$, $14\,245$ and $14\,580$ cm^{-1} in aqueous solution, shifting toward lower energy in deuterium oxide (0.19%), methanol (0.5%), nitromethane (1.2%) etc.[85], and the deuteration and solvent effects on luminescence (close to $13\,280$ cm^{-1}) were investigated[86].

For our point of view, it can be considered likely that any non-cubic distortions in glasses are indeed weak, only providing a broadening both of the quartet and doublet levels (the least in the luminescent ^2E) and that it would be very hard to conclude backwards what kind (if not several kinds) of distortion occur in a definite glass or micro-crystallites[26] in a glass-ceramic (subject to considerable local strain, when compared to large crystals of the same composition). Also, the absorption and luminescence spectra present very little clear-cut evidence for any deviations from orthoaxiality, though Eriksen and Mønsted[87] argue that the cis-complex of a rather rigid quadridentate cyclic amine and two chloride ligands is not strictly orthoaxial. One should remember[49] that glasses are prepared as mobile liquids at a quite high temperature. By cooling, a meta-stable situation is achieved at a temperature well above $200\,°C$ where the Cr-O internuclear distances hardly change anymore over time-scales of days or years. This means that the spectroscopic properties continue to be determined by the higher dispersion of internuclear distances, concomitant stronger and broader $3d^3$ absorption bands, and other characteristics of an apparent temperature much above that of the spectroscopic measurement.

Crystalline materials containing chromium(III) are known at lower temperatures to show minute effects of deviations from cubic symmetry. Thus, the separation 29 cm^{-1} of ^2E in two lines[42] in highly dilute ruby (x small) $Cr_xAl_{2-x}O_3$ is a combined higher-order effect of the prevailing trigonal symmetry (C_3 with holohedrized symmetry S_6 also called C_{3i}) and of spin-orbit coupling. The groundstate $^4A_2$ is split ten times less, as also known from zero-field splittings of electron spin resonance. Below 107 K, traces of chromium(III) in the cubic perovskite $SrTiO_3$ show ^2E split into two lines, at 4 K: $12\,597$ and $12\,600$ cm^{-1}[88]. Bergin et al.[136] studied glass-ceramics (of mullite type) obtaining fluorescent line-narrowing of ^2E between 11 and 160 K.

When the chromium(III) concentration is not low, many lines can appear in the ^2E emission, ascribed to two or more anti-ferromagnetically coupled $3d^3$ systems situated at various geometrical distributions. We already mentioned such effects observed in high-resolution luminescence spectra of glasses and glass-ceramics. The most clear-cut measurements to explain are cases of dimeric chromium(III) complexes held together (most often by two hydroxo bridges) in known crystal structures. The (rather unique) classical case[137] is the "blue basic rhodo ion" $(H_3N)_5CrOCr(NH_3)_5^{+4}$ having $S = 0$ as groundstate, followed by $S = 1$, 2 and 3 at higher energies, well described by $JS(S + 1)/2$ where the Heisenberg parameter J is 450 cm^{-1}. Hence, salts of this cation at room temperature ($kT = 210$ cm^{-1}) taking into account the number of states $(2S + 1)$ in the Boltzmann population, consists of 73.5% singlet and 25.9% triplet species, and goes diamagnetic by sufficient cooling. A large number of other dimers[89] vary J over quite a wide range, from 3 to 83 cm^{-1} (the low J values are usually detected by measuring the magnetic susceptibiliy at low temperatures). In the limit of almost paramagnetic behaviour ($kT \gg J$) the average value of $S(S + 1)$ for a system of N d^3 ions is 15 N/4 showing the additivity of each Cr(III) (Chapt. 24 of Ref. 40). The group of Hans Güdel[90] has performed detailed spectroscopic studies of such systems at low temperatures. The green corundum-type Cr_2O_3 is anti-ferromagnetic with a Néel temperature 307 K ($kT = 213$ cm^{-1}). In absorption, ^2E is a group of lines between 13 740 and 13 970 cm^{-1} [42, 91, 92]. The reason why such lines can be rather intense, is that the coupling of $S = 3/2$ ($^4A_2$) and 1/2 (^2E or $^2T_1$) provide the S values 1 and 2 of two adjacent Cr(III), included among the groundstate alternatives $S = 0$, 1, 2 and 3. In this sense, the spin-forbidden transitions become spin-allowed. However, one has to remember that group-theoretical selction rules are permissions rather than requirements. There is hardly a perceptible intensification of spin-forbidden transitions in reasonably dilute, paramagnetic systems.

The application of ruby $Cr_xAl_{2-x}O_3$ as laser provoked much interest in the (partly deleterious) effects of dimers, trimers and various larger clusters. It seems established that their concentration increases more rapidly with x than a statistical distribution on aluminium sites. This may to some extent be a question of thermodynamical equilibrium (even at the high temperature at which ruby is synthesized) because Cr(III) has a so much larger inherent ionic radius than Al(III) that its presence stores a lot of excess free energy. This excess may be less than twice or thrice as large, if shared by two or three Cr(III) (much like small mercury globules coalesce to larger drops, because of surface tension). The anti-ferromagnetic stabilization seems small, compared to kT, since the J of Heisenberg is 120 cm^{-1} [138, 139]. This may be compared with Cr(III) in a phosphate glass with an average J close to 24 cm^{-1} [94] and a value about 30 cm^{-1} in the spinel $ZnCr_xGa_{2-x}O_4$ [93]. A quite different aspect is the possible non-equilibrium clustering occurring under the preparation of ruby or glass-ceramics.

9.c Potential Hypersurfaces for Quartet and Doublet States

As soon as the quartet-quartet luminescence of chromium(III) compounds was discovered by Porter and Schläfer[54, 65, 95] the criterion for it being feasible (it may need low temperature and small chromium concentration, as discussed in 9.d, and it may still be undetectable[69] for obscure reasons) was recognized to be a short-lived $^4T_2$ being trapped at sufficiently low energy below the longer-lived ^2E. The ^2E emission is sometimes called phosphorescence, in analogy to the triplet states of organic molecules emitting to their singlet groundstate with very long half-life. Though $^4T_2$ is short-lived, it is still in many cases a *thexi* [thermally equilibrated (as far vibrations go) excited electronic] state in the sense of Adamson[96] and its thermalized form may have a few thousand cm^{-1} lower energy than the maximum of the broad absorption band $^4A_2$ to $^4T_2$. Empirically[95] it turns out that if ^2E (narrow absorption band) is 1700 to 2000 cm^{-1} below the $^4T_2$ maximum, both $^4T_2$ and ^2E emission can be observed (generally at 90 K). If the ^2E band is at least 3000 cm^{-1} below the $^4T_2$ maximum, only ^2E emission is detected. The opposite situation was missing[95] a few examples of close coincidence; if ^2E is on the high-energy side of the $^4T_2$ maximum (as in many binary and ternary halides) the only luminescence having any chance of being detected is from $^4T_2$.

Since the $^4T_2$ emission band is (at least) as broad as the absorption band, a diagram similar to Fig. 2 was proposed[54, 95] with the Cr-X internuclear distance R as variable. Condensed matter is different from most gaseous molecules by the electronic origin of the broad absorption band being difficult or impossible to locate. Since the Gaussian error-curve usually is a good approximation[38] it is likely that the origin ("zero-zero line") cannot be less than twice the one-sided half-width δ below the wave-number of the maximum (the molar extinction coefficient ε should be $\varepsilon_{max}/16$ at this point), and 2.5 δ to the origin produces $\varepsilon_{max}/87$ at the origin. Dilute ruby at low temperature[42] shows an origin at 16 767 cm^{-1} (about 1350 cm^{-1} below the maximum) followed by a vibrational structure. This origin is 2330 cm^{-1} above ^2E, precluding $^4T_2$ emission.

A gaseous molecule MX_6 such as IrF_6 has $15 = 3(7 - 2)$ mutually independent internuclear distances among the $7 \cdot 6/2 = 21$ occurring. Assuming the approximation of Born and Oppenheimer[97] of factorization of the total wave-function in a translational, a rotational, a vibrational and an electronic part, the 15 degrees of freedom of nuclear motion provide a potential surface (energy as a function of nuclear positions) in a 16-dimensional space. What kineticists call the "reaction path" (if not the "reaction coordinate") is the most favorable connection between two different nuclear constellations, but it still takes place in 16 dimensions. When the totally symmetric "breathing" mode (multiplying all internuclear distances by the same scaling factor close to 1) is so popular among spectroscopists, it is not only because a diatomic molecule MX has only one internuclear distance (providing a potential curve in two dimensions) but also because the promotion of electrons to anti-bonding orbitals [or their energetically favorable demotion, on the condition of accompanying increased S.C.S. interelectronic repulsion, as known[98] from the first excited quartet states of manganese(II)] can be represented on such a totally symmetric coordinate, e.g. R if all the ligands are identical. It has little relevance to discuss whether octahedral MX_6 incorporated in a solid has a 22- rather than 16-dimensional potential surface, but the thexi state[96] achieves a vibrational wave-function very close to a (relative) minimum of this potential surface. Hence, luminescence may sometimes occur at a much lower energy than predicted by a totally symmetric R variation.

For the chemist, the *Jahn-Teller effect* is, in a way, the obverse side of the Tanabe-Kamimura theorem[79]. The abstract formulation of the Jahn-Teller theorem is that two or more electronic states having identical energy in a definite symmetry of a non-linear molecule or polyatomic ion (with exception of Kramers doublets in the case of an odd number of electrons) spontaneously separate in energy, such that (at least) the state of lowest energy has distorted the prevailing symmetry. Actually, it is almost excluded to detect Jahn-Teller effect in lanthanide compounds[50] if some of the $(2J + 1)$ states of the lowest J-level have the same, marginally lower energy. On the other hand, in the d groups, the sub-shell M.O. configurations are an excellent indicator of perceptible Jahn-Teller effect. If two or three d-like orbitals have exactly the same one-electron energy in a given high symmetry of the molecule or complex ion, and if these orbitals are occupied in unbalanced way (by one and by no electron; or by two and one electron) the nuclear positions are modified in such directions that the orbital carrying more (or one) electrons is stabilized relative to the orbital carrying one electron less. Even in the case of one electron remaining in the latter orbital, it is still (within certain limits) an energetic advantage, though the second orbital may be more destabilized than the first is stabilized. There are no such limits imposed, if the second orbital is empty. On the other hand, if two or three equivalent d-like orbitals carry one electron each, there is no first-order energetic advantage of distorting the symmetry, according to the Tanabe-Kamimura theorem, and the general environment (which determines the equilibrium symmetry in the first place) then counteracts spontaneous distortions. Hence, the strongest Jahn-Teller distortions are expected in octahedral d-group compounds if the occupation numbers differ of the σ-anti-bonding orbitals $(3z^2 - r^2)$ and $(x^2 - y^2)$. This is the case[99, 100] for the $3d^4$ chromium(II) and manganese(III), and $3d^9$ copper(II) compounds. The quantitative extent of the Jahn-Teller distortion can be treated[101] by the angular overlap model. It is interesting to note that strong Jahn-Teller effect is not restricted to copper(II) which might be suspected[102, 103] for higher one-shot ionization energy of several of the nine d-like electrons than of the filled ligand orbitals, but also occurs for the reducing chromium(II). The comparison of spectroscopic and crystallographic properties of many copper(II) compounds[38, 40, 56, 100, 104] allow to realize the importance of differing timescales of the various techniques of measurement. In view of the non-negligible π-anti-bonding effects of anion ligands[77], it is not surprising that a relatively weak Jahn-Teller effect is induced by unbalanced occupation of π-anti-bonding orbitals, e.g. (xz; yz; xy) in $3d^1$ titanium(III)[44].

It is not always realized that excited states may be strongly influenced by the Jahn-Teller effect, though this is not the only reason for a large Stokes shift between the electronic origin of the absorption band, and the maximum of the broad emission band corresponding to the excited state, as known for many post-transitional elements[50, 105] such as bismuth(III). Frequently, the Stokes threshold cannot be detected, because the emission occurs at wave-numbers far below the beginning of absorption, such as the rather extreme case of the $3d^6$ cobalt(III) cyanide complex $Co(CN)_6^{-3}$ having the maximum of the absorption band corresponding to the first excited singlet$^1T_1$ at $32\,100$ cm^{-1} and the nine states belonging to $^3T_1$ not far from $26\,000$ cm^{-1}. Nevertheless, a broad emission to the groundstate $^1A_1$ has a maximum at $14\,200$ cm^{-1}[106, 107] with life-time 0.7 ms at 77 K and 0.008 ms at 300 K. The transient $^3T_1$ measured at 94 K in an organic glass[108] has three quite strong absorption bands at $21\,600$, $23\,800$ and $26\,000$ cm^{-1} ascribed to levels of $(t_2)^4 (e)^2$, though the assumed ratio $C/B = 7.92$ is unlikely high[20]. Both groups of authors[107, 108] expect totally symmetric expansion of the thexi state $^3T_1$ (to the extent[108] of 0.35 Å) but it is by no means certain that the equilibrium geometry of $^3T_1$ is close to regular octahedral. The concentration quenching and energy transfer in mixed crystals $A_3[Cr(CN)_6]_x[Co(CN)_6]_{1-x}$ where A is various univalent cations[109, 110] or silver(I)[111] have been thoroughly studied. $Ag_3Co(CN)_6$ has bent CoCNAgNCCo bridges[112]. The emission and reflection spectra of ten different anhydrous and hydrated $A_3Cr(CN)_6$ were compared[113].

Such observations as in hexacyanocobaltate(III) of luminescence at strongly decreased energy usually have problems of being accepted by the referee. In nearly all cases, a tiny impurity takes over the energy (for many years, one could buy two qualities of anthracene, a very expensive fluorescing violet, and one fluorescing green due to a few parts per million of tetracene). However, $^3T_1$ of octahedral symmetry belongs to the sub-shell configuration $(t_2)^5$ (e) with one strongly anti-bonding electron. By a mechanism closely related to the Jahn-Teller effect, the excited state has the time to rearrange to some kind of distorted trigonal prism, or possibly a hedgehog-like structure, having two orbitals filled by four d-like electrons, and two higher orbitals (of identical or similar energy) each containing one d-like electron. This thexi "isomer" has a triplet ground-state. The subsequent broad-band emission showing a Stokes shift of order $12\,000$ cm^{-1} is strongly prevented by the Franck-Condon principle from being observed as an absorption band in the orthoaxial groundstate, besides its spin-forbidden character.

Among the chromium(III) octahedra discussed here, the ^2E emission band is uniformly narrow, showing that the vibrational extension of the ^2E thexi state is "projected" down on the part of the $^4A_2$ potential surface running almost strictly parallel with the ^2E potentialsurface (in spite of both being U-levels in O_h[58, 114] and hence containing two separable Kramers doublets each). The somewhat broader emission[80, 81] of the lowest component of $^2T_1$ in tetragonal compounds corresponds to less perfectly parallel potential surfaces. One may discuss[40] whether strong second-order separation of the two components of ^1E in $3d^8$ nickel(II) normally keeping close together (roughly regular octahedral NiX_6) because of the Tanabe-Kamimura theorem, but providing a sub-shell configuration in D_{4h} $(b_{2g})^2$ $(e_g)^4$ $(a_{1g})^2$ having as only state the singlet groundstate of quadratic NiX_4 [with the other Tanabe-Kamimura $(b_{2g})^2$ $(e_g)^4$ $(b_{1g})^2$ at far higher energy] is not an indirect (sort of metaphorical) example of the Jahn-Teller effect (though not of the Jahn-Teller theorem). The difference from the $^3A_2$ groundstate of NiX_6 is that *two* electrons in the same a_{1g} ($3z^2 - r^2$) orbital can be accommodated, on the condition of diamagnetism, and then binding only four ligands strongly in the xy-plane, which do not need to worry about making the empty b_{1g} ($x^2 - y^2$) orbital heavily anti-bonding, though it shifts the first spin-allowed absorption bands of such orange or yellow complexes up above $20\,000$ cm^{-1}.

Orgel[115] was the first to rationalize the colour difference between red ruby $Cr_xAl_{2-x}O_3$ and the isotypic green Cr_2O_3 (and other green materials, such as emerald $Cr_xAl_{2-x}Be_3Si_6O_{18}$, chromium(III) containing glasses, etc.) all having relatively similar absorption spectra, as a variation of Δ, increasing dramatically for shorter R. We still do not know exactly to what extent the squeezing of Cr(III) into dilute (pink) ruby makes it adopt the low R characterizing pure corundum Al_2O_3 or a compromise[115] is reached, where the six closest oxygen nuclei are pushed away, though not nearly as much as in stoichiometric Cr_2O_3. Like in fluorite-type CeO_2 this must be determined by the strong repulsion between adjacent oxide having exceptionally short internuclear distances. With exception of the alexandrite C_i site discussed below (9d) the Δ values of Cr(III) in mixed oxides vary between 1.05 and 0.93 times $\Delta = 17\,450$ cm^{-1} for the aqua ion, and in oxide glasses down to 0.86. This is a rather modest variation, compared[20] to this ratio (1.02 for NiO) in mixed nickel(II) oxides varying down to 0.70 for ilmenite-type $Ni_xCd_{1-x}TiO_3$ and 0.56 for the (elpasolite?) superstructure of perovskite $Ba_2Ca_{1-x}Ni_xTeO_6$. One conclusion may be that Cr(III) decides R to a better precision than Ni(II) more readily adapting to large cadmium (II) and calcium(II) sites.

As far as the $^4T_2$ potential surface goes, by far the simplest picture would be totally symmetric expansion, neglecting all the other normal modes. In such a model, it becomes legitimate to speak about "low-field" complexes[65, 95] with (ideally the origin of) $^4T_2$ below ^2E, and "high-field" cases with ^2E below $^4T_2$. The distance (having no problems of virbrational co-excitation) from the groundstate $^4A_2$ to ^2E is asymptotically $21\,B$ for very high ratio $\Sigma = \Delta/B$, assuming the Racah parameter $C = 4\,B$. In the Tanabe-Sugano diagram Fig. 1, the crossing with $^4T_2$ occurs at Σ somewhere between 18 and 19. However, the B value obtained from the spin-forbidden intra-sub-shell transitions in $(t_2)^3$ is markedly larger than from the distances between $^4T_2$ and the two $^4T_1$ but distinctly smaller than $B_0 = 918$ cm^{-1} derived from ^4F and ^4P in gaseous Cr^{+3}. It is attractive (Eq. 5.25 of Ref. 20) to write

$$\beta_{55} = a_5^4\,\beta_0 \qquad \beta_{35} = a_5^2\,a_3^2\,\beta_0 \tag{4}$$

where a_5^2 is the squared 3d amplitude of t_2 orbitals (called γ_5 by Bethe) in L.C.A.O. descriptions, a_3^2 the squared 3d amplitude of e orbitals (γ_3) and β_0 the part of the nephelauxetic effect due to radial expansion (the central field not being so deep in Cr(III) as in gaseous Cr^{+3}). In absence of π-anti-bonding, a_5^2 is presumably 1, and hence, $\beta_{55} = 0.84$ in Cr(NH$_3$)$_6^{+3}$ and 0.71 in the isoelectronic (and highly reactive) Cr(CH$_3$)$_6^{-3}$ should represent β_0. Since $\beta_{35} = 0.71$ in Cr(NH$_3$)$_6^{+3}$ and 0.57 in Cr(CH$_3$)$_6^{-3}$ [almost the same as 0.58 for Cr(CN)$_6^{-3}$ whereas β_{35} is only 0.42 for Co(CN)$_6^{-3}$] the ratio 0.71/0.84 = 0.85 should indicate the a_3^2 in the ammonia complex (a fractional chromium charge close to 1.7) and 0.57/0.71 = 0.80 = a_3^2 in Cr(CH$_3$)$_6^{-3}$ which seems[20] to have a chromium charge of order 1. The β_{35} values derived from spin-allowed inter-sub-shell transitions decrease from 0.79 for Cr(OH$_2$)$_6^{+3}$ to 0.71 for ruby, 0.68 for the tris-oxalato complex, 0.56 for LaCrO$_3$ and 0.52 for Cr$_2$O$_3$ to 0.45 for Cr(S$_2$P(OC$_2$H$_5$)$_2$)$_3$. However, in spite of the complications related to Eq. (4), the choice between the lowest excited level being $^4T_2$ of $(t_2)^2(e)$ or ^2E of $(t_2)^3$ is closely similar to the choice between 3d^6 iron(II) having high-spin groundstate $^5T_2$ belonging to $(t_2)^4(e)^2$ or low-spin $^1A_1$ belonging to $(t_2)^6$. In both cases, a spin-pairing energy representing a definite combination of Racah (or S.C.S.) parameters (this concept[20] has recently been further analyzed[116–118]) such as quartet states (on the average) being $3\,D$ more stable than doublet states (and quintet states $6\,D$ more stable than singlet states) must be compared with Δ or with $2\,\Delta$, respectively. Orgel pointed out in 1956 that octahedral groundstates being high-spin ("low-field") or low-spin ("high-field") in practice show a no-man's land of Δ (or more strictly speaking of Σ values). There is a lower limit for low-spin behaviour. However, in an interval between a factual higher limit for high-spin groundstate, and this lower limit for low-spin groundstate, no examples are observed, because such high-spin cases contract their R values, increasing their Δ, and finally gain energy by becoming low-spin cases just above the lower limit.

The choice between $^4T_2$ and ^2E is, of course, not a choice of groundstate. The spectroscopic properties discussed here are, as far as absorption goes, determined by the vibrational amplitude of the $^4A_2$ groundstate. By a "vertical" projection (satisfying the Franck-Condon principle) it may hit ^2E at lowest energy (as is the case without ambiguity in ruby) or the beginning of the broad band due to $^4T_2$. The observed one-sided half-widths δ (typically 800 to 1800 cm^{-1} [38, 39]) are compatible with the force-constant of the groundstate, and the slope of the excited state[38, 40]. Hence, the width 2δ of such inter-sub-shell absorption bands being an order of magnitude larger than kT at ambient temperature is readily rationalized. It is possible to look at this question in another perspective less familiar to molecular spectroscopists. During the vibration of d^q complexes in their electronic groundstate, their Gaussian-shaped probability distribution of R centered around a mean value R_0 sweeps the Orgel diagram[98] (or assuming roughly invariant B, the Tanabe-Sugano diagram) with a Gaussian-shaped image having Δ decreasing, say 5%, for each percent increase of R relative to R_0. Hence, the excited $^4T_2$ appoaches its higher R_{eq} value by thermalization, and any luminescence from this thexi state has lower energy for two complementary reasons. One way of expression is that Δ has been multiplied by $(R/R_{eq})^{-5}$ (or whatever exponent[46-48]) and another that the final state of the luminescence leaves the electronic groundstate vibrationally highly excited. Seen from a purely spectroscopic point of view, the coexistence of long-lived ^2E and short-lived $^4T_2$ leaves the possibility open of thermal excitation of ^2E to the vibronic quasicontinuum of $^4T_2$ (the lowest vibrational state of $^4T_2$ on Fig. 2 being above the lowest vibrational state of ^2E) storing energy in ^2E and releasing it more slowly through $^4T_2$ than usually (the life-time of $^4T_2$ usually being hundreds of times shorter than of ^2E). Such thermal excitation to adjacent higher-lying levels is well-known in lanthanides[49, 50] such as the green emission[119] from $^2H_{11/2}$ of $4f^{11}$ erbium(III) at some 600 cm^{-1} higher energy than $^4S_{3/2}$; or from two (slightly enigmatic) levels of bismuth(III)[105] showing a dramatic jump of the Stokes shift as a function of passing a narrow temperature interval.

Taking all of these features into account, the $^4T_2$ emission seems qualitatively compatible with totally symmetric "breathing" with one R parameter varying, though it may seem worrying that the $^4T_2$ broad emission bands (if detected) have maxima well below 14 000 cm^{-1} (715 nm) and stretch long out in the near-infrared. If the Stokes shift was as large[107, 111] as 12 000 cm^{-1} for $^3T_1$ of Co(CN)$_6^{-3}$ the $^4T_2$ would simply degrade visible light to heat (like so many other materials not showing perceptible luminescence) and still, a Stokes shift 3000 to 4000 cm^{-1} is more than a-quarter of the wave-number of the emission maximum. These loose hints of non-totally-symmetric distrotions of the $^4T_2$ thexi state have been confirmed in one clear-cut case[120]. The cubic elpasolites Cs$_2$NaMX$_6$ (alternating strings of XNaXMXNaXM... nuclei on each Cartesian axis running parallel with the distances half the unit-cell parameter) where, around 8 K, luminescence from $^4T_2$ shows a pronounced vibrational structure, starting e.g. in Cs$_2$NaCr$_{0.01}$Y$_{0.99}$Cl$_6$ at three origins (corresponding to three marginally different sites) at 11 797, 11 752 and 11 711 cm^{-1} and stretching down to 10 000 cm^{-1}. Detailed analysis of the normal modes[120] (multiply co-excited in the emission band) demonstrates a quite strong Jahn-Teller effect, the instantaneous picture of $(t_2)^2$ (e) having elongated four Cr-X distances in one plane by 0.09 Å in Cs$_2$NaCr$_x$In$_{1-x}$Cl$_6$, by 0.10 Å in Cs$_2$NaCr$_x$Y$_{1-x}$Cl$_6$ and by 0.15 Å in Cs$_2$NaCr$_x$Y$_{1-x}$Br$_6$ (relative to the $^4A_2$ groundstate) and shortened the two perpendicular Cr-X by about a-quarter of these amounts. Said in other words, the Jahn-Teller effect operates in the opposite direction of $3d^9$ copper(II). In these crystals, ^2E (situated at higher energy than the $^4T_2$ absorption maximum) does not contribute to the luminescence. The life-time of $^4T_2$ is quite long at T below 200 K, in the 0.04 to 0.15 ms range, and close to the radiative life-time.

There is another case, which suggests insufficiency of the (one R)-model, if taken at face value. The cubic elpasolite K$_2$NaCr$_{0.05}$Ga$_{0.95}$F$_6$ has a zero-phonon line (at 4 K) at 15 041 cm^{-1} followed by 52 vibrational components in emission having an average position about $14\,000$ cm^{-1} [55]. The origin can only be ascribed to the lowest U level being mainly ^2E, but possibly containing some $^4T_2$ squared amplitude [58, 114]. The life-time [55] is 0.28 ms at 300 K, 0.53 ms at 77 K and 0.62 ms both at 21 and 14 K. This is perhaps not as long as expected for ^2E, but the curious fact is the long vibronic progression. Admittedly, $(NH_4)_3CrF_6$ is not the same compound (though both contain CrF_6^{-3} groups) but it shows [65] at 87 K a broad, asymmetric emission band with maximum $12\,800$ cm^{-1} and half-width δ toward lower energy 1300 cm^{-1} typical for $^4T_2$ emission. At room temperature the reflection spectrum shows two moderately narrow peaks at $15\,060$ and $15\,670$ cm^{-1} (and a shoulder at $16\,500$ cm^{-1}) superposed a broad band with the mean position $15\,600$ cm^{-1}.

Though it is interesting [121] that intermediate coupling calculations in O_h for CrF_6^{-3} ($\Delta = 16\,100$ cm^{-1}) provide doublet-like levels (in cm^{-1}) U $14\,586$; U $15\,296$; E' $15\,345$, and four $^4T_2$ levels E'' $16\,032$; U $16\,077$; E' $16\,165$ and U $16\,174$, followed by the next doublet U $21\,773$ and E'' $21\,858$, it is perfectly clear today that such a treatment cannot be carried out the same way as for an atomic spectrum. The minimum requirement is an intrinsic band-width proportional to the $(t_2)^2(e)$ character of the excited level [66, 98] and a nagging doubt is that one may have to take 16-dimensional potential surfaces into account.

9.d Yields of Luminescence

The technological purpose of many recent studies of chromium(III) $^4T_2$ near-infrared luminescence in crystals or glasses is the operation of *tunable lasers* [122] for which titanium(III) in $Ti_xAl_{2-x}O_3$ has also been proposed [44]. The necessary (but not sufficient) condition for laser action [50] is population inversion, that the concentration of the initial state of the luminescent transition is higher than of the final state. Though the ruby laser (invented 1960) is a three-level laser (the final state is the groundstate $^4A_2$) it is much easier to achieve population inversion in a four-level laser, where the luminescence terminates at a level at least, say 5 kT, above the groundstate. Tunable lasers (using a vibronic quasi-continuum) are intrinsically four-level lasers, and are mostly known in the laboratory as dye lasers [13] were some 20 organic colorants (e.g. rhodamine G) can be selected, to emit strictly monochromatic light (inside a broad emission band) anywhere in the whole visible range. Solid-state tunable lasers [122] are intended also to achieve high power levels, though at present, the terawatt lasers [123, 124] emitting 10^{-9} s pulses, each carrying several thousand joules of energy, are silicate glass using the $4f^3$ neodymium(III) $^4F_{3/2} \rightarrow {}^4I_{11/2}$ (final state 2000 cm^{-1} above the groundstate). Such terawatt lasers allow thermonuclear fusion in deuterium-tritium mixtures.

Chromium(III) luminescence presents many of the (somewhat unpredictable) aspects of luminescence outside the lanthanides [50]. Traces can be detected, such as the red line of cathodo- and photo-luminescence of apparently pure Al_2O_3 ascribed by Crookes [125] to a rare earth (since the line is so sharp) and that Lecoq de Boisbaudran (who discovered gallium, samarium and dysprosium) argued is due to chromium. A typical case are mixed crystals [126] of acetylacetonates (Cr aca$_3$)$_x$ (Al aca$_3$)$_{1-x}$ having the ^2E life-time 0.43 ms between $-170\,°C$ and $-90\,°C$ for x = 0.001, and then rapidly vanishes by warming up to $-50\,°C$. For x = 0.1, the quenching starts at slightly lower T, and x = 0.5 has the life-time decreasing from 0.2 ms at low T, and vanishing at $-100\,°C$. Kinetic arguments were given[126)] for $^2T_1$ acting as an intermediate state. The dramatic dependence of Cr(III) life-time, especially of $^4T_2$, on temperature was early recognized[54, 65, 95)] and activation energies determined for thermal quenching. This shortening of life-time is much less pronounced[55)] for $K_2NaCr_{0.05}Ga_{0.95}F_6$ than for undiluted K_2NaCrF_6. Quite generally, it is difficult to analyze the intricate synergism between thermal and concentration quenching. There seems to be a certain similarity between the thermal quenching of europium(III) and other $4f^q$ J-level luminescence by the broad electron transfer bands at much higher "vertical" energies[127)] providing potential surfaces, of which the description should not be restricted to totally symmetric stretching, but needs the multi-dimensional aspects. As first pointed out by Weber[128)] the J-levels have multi-phonon non-radiative de-excitation determined by the ratio between the highest characteristic vibrational frequency and the energy gap between the luminescent J-level and the closest lower J-level[49, 50)]. This mechanism is not predominantly dependent on T, but can be demonstrated, also in chromium(III) complexes[71, 82, 96)] by deuteriation. $Cr(OD_2)_6^{+3}$ in the alum $KAl(SO_4)_2$, 12 D_2O shows sharp ^2E emission lines[129, 130)] with life-time τ decreasing from 0,74 ms at 4 K to 0.047 ms at 77 K[130)]; or $\tau = 0.13$ ms with $\eta = 0.001$[129)] accompanied by weaker $^4T_2$ broad-band luminescence with η close to 10^{-4}.

There is no doubt that the ordinary process of quenching in $^4T_2$ is not related to multi-phonon relaxation, but is enhanced by increasing chromium(III) concentration, as already furthered by anti-ferromagnetic coupling between two or more $^4A_2$ systems. In glasses, it is typical[131)] that 0.03 to 0.1 mole% chromium(III) in a calcium borate glass (emitting both from ^2E and $^4T_2$) has the 800 nm broad-band $^4T_2$ emission decreasing its quantum yield from 0.25 below 122 K to 0.01 at room temperature. It was checked that no "killer centers" like traces of iron(III) or iron(II) were influential. At 300 K, the highest η known of $^4T_2$ emission from a conventional glass is 0.23 in a lithium lanthanum phosphate glass[12)] containing $1.4 \cdot 10^{19}$ Cr/cm^3 or 0.023 mole/dm³. Already at twice as high Cr(III) concentration, η is down at 0.16, and for 0.14 mole Cr/dm^3, η is only 0.04.

Though glass-ceramics have distinctly higher η in many cases, it is worthwhile to compare with *alexandrite* crystals $Cr_xAl_{2-x}BeO_4$ being one of the most popular solid-state tunable lasers[132, 133)] recently studied by elaborate four-wave-mixing techniques. Chrysoberyl Al_2BeO_4 is isotypic with the mineral olivine $Mg_{2-x}Fe_xSiO_4$ (and other orthorhombic crystals, such as Ni_2SiO_4 and Mg_2GeO_4) and hence is a (tetrahedral) beryllate in spite of the chemical formula analogous to spinel. There are two octahedral sites, one with a centre of inversion C_i and one with a mirror plane C_s (and the holohedrized symmetry C_{2h}). Only 22% of the chromium (in a crystal with a total concentration $1.15 \cdot 10^{19}$ Cr/cm^3 or 0.019 mole/dm³) was found on the C_i site, and 78% on C_s sites. Δ on the C_i site is well above 20 000 cm^{-1} (and possibly not much below 21 550 cm^{-1} of $Cr(NH_3)_6^{+3}$). When the crystal is excited at 488 nm, two very weak, sharp ^2E lines[132)] occur at 690 and 697 nm, whereas excitation at longer wave-length only provides (C_s) lines at 678 and 681 nm. This shift about 300 cm^{-1} suggests shorter Cr-O distances on the C_i site, with a smaller β_0 in Eq. (4). There is some evidence that the Mg-O distances in Mg_2SiO_4 differ as much as 2.14 Å for the C_s and 2.10 Å for the C_i site. It might, of course, be suggested that Δ is large on the C_i site because the π-anti-bonding effect on t_2 orbitals[77)] is exceptionally small, but then, a_5^2 in Eq. (4) should have been closer to 1. Anyhow, the ^2E lines from the C_i site show the very long life-time 63 ms. The broad absorption band corresponding to the $^4T_2$ states (in O_h) is highly dependent on the direction of polarized light[132)] much like in hexagonal beryl-type emerald $Cr_xAl_{2-x}Be_3Si_6O_{18}$. In both cases, the strong mono-axial distortion also splits ^2E more than it does in ruby. The measured life-time of ^2E on the C_s site is only 2.4 ms at 6 K[133] but this is taking into account the competing loss of energy by energy transfer, or by thermal excitation, to the $^4T_2$ (of which the origin is said to occur 800 cm^{-1} above ^2E). Long-range energy migration of ^2E excitation between C_s sites becomes important below 150 K. At 300 K, the thexi ^2E state is Boltzmann-depopulated to $^4T_2$ to an extent about e^{-4} (assuming two Kramers doublets left at low energy by the mono-axial perturbation of $^4T_2$) or about 1.8%; or 5% if the non-cubic influence is negligible. In these two cases, the life-time (and η) of ^2E goes down to the half, if the observed (radiative *and* non-radiative) life-time of $^4T_2$ is 54 or 18 times shorter than of ^2E, respectively. If the non-radiative de-excitation of $^4T_2$ is weak, η of the broad band emission then becomes almost 0.5.

9.e Pragmatic Conclusions

In many cases of broad-band $^4T_2$ luminescence of chromium(III), the major drawback is that the life-time τ and the quantum yield η (the ratio τ/η is, in principle, invariant[49, 50] and represents the Einstein radiative life-time) within a narrow temperature interval decrease from 90% of the asymptotic values at low temperature to 5% or less. This pronounced thermal quenching (and to some extent also the concentration quenching which may occur even at less than 0.05 mole/dm^3) is far more unpredictable than quenching of lanthanide luminescence, and quite difficult to rationalize at the moment. However, one may add the constructive argument that new preparative techniques, and ameliorated understanding, might bring $^4T_2$ lasers even at 300 or 400 K. Though it needs much higher threshold levels of pumping, $\eta = 0.2$ might still suffice, though $\eta = 0.5$ obviously is preferable. Further on, the lasing threshold can be significantly lowered by energy transfer[140] from chromium(III) to resonant or lower-lying J-levels of several lanthanides, such as[141] neodymium(III).

At room temperature, several glass-ceramics show η quite close to 1 for the broad-band $^4T_2$ emission, and distinctly far higher than all known values for homogeneous glasses of mixed oxides or fluorides. The general image connected with glass-ceramics is the opalescent shock-resistant kitchenware; but it is possible to make the micro-crystallites so much smaller than the wave-length of violet light that light-scattering is not a prohibitive problem for laser action. Compared to the orthorhombic crystals alexandrite (one of the most promising tunable solid-state lasers[132, 133]) glass-ceramics are isotropic, and homogeneous on a sufficiently small scale of length, that cracking due to violent thermal fluctuations and to high-amplitude standing waves is less of a problem than for non-cubic crystals. The compositions of glass-ceramics that may still be prepared, reminds one of Edison's 50 000 experiments to make filaments for electric lamps, but on the other hand, emission in the near-infrared did not attract any attention on a large scale before the neodymium(III) vitreous and crystalline lasers. Micro-crystalline glass-ceramics may very well be the solution to a flock of problems encumbering crystalline laser materials.

In addition to "ligand field" theory and group-theoretical engineering[18, 20, 40, 73–78] rationalizing the excited levels of chromium(III) since about 1954, it is important to remember the extensive studies by Neuhaus[134] of absorption spectra of minerals coloured by this 3d^3 system, such as ruby, emerald, the cubic garnet-type uvarovite, the monoclinic muscovite (mica)-type fuchsite, etc. These studies were continued with carefully designed, mixed oxides at the University of Bonn by O. Schmitz-DuMont and Reinen[43] (now at the University of Marburg). Minerals have presented us with opportunities to conceive new ideas, we would never have got otherwise.

Acknowledgements. The authors are grateful to Drs. A. Buch and M. Ish-Shalom for skilful preparation of glass-ceramics; to Dr. A. Kisilev, and subsequently Dr. M. Eyal and Dr. Y. Kalisky, for their thorough experimental studies; to Professor G. Boulon, Université de Lyon I, for pleasant collaboration on time-resolved and line-narrowing spectroscopy; to Professor C. E. Schäffer, University of Copenhagen, for fruitful discussions on chromium(III) energy levels and the angular overlap model; and finally to Mrs. E. Greenberg for her assistance in preparing the manuscript. The experimental work carried out in Israel was supported by the US Army Contract no. DAJA 45-85-C-0051.

10 References

1. Reisfeld, R., Kisilev, A., Greenberg, E., Buch, A., Ish-Shalom, M.: Chem. Phys. Lett. *104,* 153 (1984)
2. Poncon, V., Bouderbala, M., Boulon, G., Lejus, A.-M., Reisfeld, R., Buch, A., Ish-Shalom, M.: ibid. *130,* 444 (1986)
3. Kisilev, A., Reisfeld, R., Greenberg, E., Buch, A., Ish-Shalom, M.: ibid. *105,* 405 (1984)
4. Reisfeld, R., Kisilev, A., Buch, A., Ish-Shalom, M.: ibid. *129,* 446 (1986)
5. Poncon, V., Kalisky, J., Boulon, G., Reisfeld, R.: ibid. *133,* 363 (1987)
6. Reisfeld, R.: Potential uses of Cr(III) doped transparent glass-ceramics in luminescent solar concentrators. Report Series Swedish Academy of Engineering Sciences in Finland (Helsinki) *40,* part I (Proc. Advanced Summer School on the Electronic Structure of New Materials, Loviisa, August 1984) pp. 7–34, Helsinki 1985
7. Reisfeld, R.: Materials Science and Engineering *71,* 375 (1985)
8. Buch, A., Ish-Shalom, M., Reisfeld, R., Kisilev, A., Greenberg, E.: ibid. *71,* 383 (1985)
9. Reisfeld, R., Kisilev, A., Jørgensen, C. K., Buch, A., Ish-Shalom, M.: Quartet-quartet luminescence of chromium(III) in glass-ceramics. Abstracts 167. Meeting of Electrochem. Soc. on Luminescent and Display Materials Group, Toronto, May 1985
10. Durville, F., Champagnon, B., Duval, E., Boulon, G., Gaume, F., Wright, A. F., Fitch, A. N.: Phys. Chem. Glasses *25,* 126 (1984)
11. Kisilev, A., Reisfeld, R.: Solar Energy *33,* 163 (1984)
12. Reisfeld, R., Kisilev, A.: Chem. Phys. Lett. *115,* 457 (1985)
13. Schäfer, F. P. (ed.): Dye Lasers, Springer-Verlag, Berlin/Heidelberg/New York 1977
14. Boyd, R. N., Owen, J. F., Teegarten, K. T.: IEEE J. Quantum Electron. *14,* 697 (1978)
15. Walling, J. C. Jenssen, H. P., Morris, R. C., O'Dell, E. W., Peterson, O. G.: Opt. Letters *4,* 182 (1979)
16. Struve, B., Huber, G., Laptev, V. V., Shcherbakov, L. A., Zharikov, E. V.: Appl. Phys. *B30,* 117 (1983)
17. Drube, J., Struve, B., Huber, G.: Opt. Commun. *50,* 45 (1984)
18. Sugano, S., Tanabe, Y., Kamimura, H.: Multiplets of Transition Metal Ions in Crystals, Academic Press, New York 1970
19. Reisfeld, R., Jørgensen, C. K.: Cr(III) broad-band quartet-quartet luminescence in glass-ceramics with higher yield than in complexes. King, E. L., Busch, D. H., Sievers, R. E. (eds.), Abstracts of 23. Int. Conf. Coordination Chemistry, Boulder CO 1984, p. 162
20. Jørgensen, C. K.: Oxidation Numbers and Oxidation States, Springer-Verlag, Berlin/Heidelberg/New York 1969
21. Kenyon, P. T., Andrews, L., McCollum, B. C., Lempicki, A.: IEEE J. Quantum Electron. *18,* 1189 (1982)

22. Kalisky, Y., Poncon, V., Boulon, G., Reisfeld, R., Buch, A., Ish-Shalom, M.: Chem. Phys. Lett. *136*, 368 (1987)
23. Kisilev, A., Reisfeld, R., Buch, A., Ish-Shalom, M.: ibid. *129*, 450 (1986)
24. Bouderbala, M., Boulon, G., Lejus, A.-M., Kisilev, A., Reisfeld, R., Buch, A., Ish-Shalom, M.: ibid. *130*, 438 (1986)
25. Boulon, G.: Materials Chemistry and Physics *16*, 301 (1987)
26. Reisfeld, R., Kisilev, A., Buch, A., Ish-Shalom, M.: J. Noncryst. Solids *91*, 333 (1987)
27. Wang, M. L., Stevens, R., Knott, P.: Glass Technology *23*, 238 (1982)
28. Durville, F., Champagnon, B., Duval, E., Boulon, G.: J. Phys. Chem. Solids *46*, 701 (1985)
29. Bouderbala, M., Boulon, G., Reisfeld, R., Buch, A., Ish-Shalom, M., Lejus, A.-M.: Chem. Phys. Lett. *121*, 535 (1985)
30. Derkosch, J., Mikenda, W., Preisinger, A.: Spectrochim. Acta *32 A*, 1759 (1976)
31. Mikenda, W., Preisinger, A.: J. Luminescence *26*, 53 and 67 (1981)
32. Mikenda, W.: ibid. *26*, 85 (1981)
33. Derkosch, J., Mikenda, W.: ibid. *28*, 431 (1983)
34. Imbusch, G. F. in: Energy Transfer Processes in Condensed Matter (ed. DiBartolo, B.) *114*, 47 (1984), Plenum Press, New York
35. Landry, R. J., Fournier, J. T., Young, C. G.: J. Chem. Phys. *46*, 1285 (1967)
36. Clarke, R. H., Andrews, L. J., Frank, H. A.: Chem. Phys. Lett. *85*, 161 (1982)
37. Abdrakhmanov, R. S., Konoval, E. M., Shishkin, I. V.: Fizika i Khimiya Stekla *9*, 403 (1983)
38. Jørgensen, C. K.: Absorption Spectra and Chemical Bonding in Complexes, Pergamon, Oxford 1962, 2. ed. 1964
39. Jørgensen, C. K.: Adv. Chem. Phys. *5*, 33 (1963)
40. Jørgensen, C. K.: Modern Aspects of Ligand Field Theory, North-Holland, Amsterdam 1971
41. McClure, D. S.: Solid State Phys. *9*, 399 (1959)
42. McClure, D. S.: J. Chem. Phys. *36*, 2757 (1962); *38*, 2289 (1963)
43. Reinen, D.: Structure and Bonding *6*, 30 (1969)
44. Reisfeld, R., Eyal, M., Jørgensen, C. K.: Chimia *41*, 117 (1987)
45. Reisfeld, R., Eyal, M., Jørgensen, C. K., Guenther, A. H., Bendow, B.: ibid. *40*, 403 (1986)
46. Hitchman, M. A., Waite, T. D.: Inorg. Chem. *15*, 2150 (1976)
47. Smith, D. W.: Structure and Bonding *35*, 87 (1978)
48. Jørgensen, C. K., Faucher, M., Garcia, D.: Chem. Phys. Lett. *128*, 250 (1986)
49. Reisfeld, R., Jørgensen, C. K.: Handbook on the Physics and Chemistry of Rare Earths (eds. Gschneidner, K. A., Eyring, L.) *9*, chapter 58, pp. 1–90, North-Holland, Amsterdam 1987
50. Reisfeld, R., Jørgensen, C. K.: Lasers and Excited States of Rare Earths, Springer-Verlag, Berlin/Heidelberg/New York 1977
51. Jørgensen, C. K., Reisfeld, R.: Structure and Bonding *50*, 121 (1982)
52. Jørgensen, C. K.: Acta Chem. Scand. *8*, 1495 (1954)
53. Schäffer, C. E.: J. Inorg. Nucl. Chem. *8*, 149 (1958)
54. Porter, G. B., Schläfer, H. L.: Z. Physik. Chem. *37*, 109 (1963)
55. Ferguson, J., Guggenheim, H. J., Wood, D. L.: J. Chem. Phys. *54*, 504 (1971)
56. Jørgensen, C. K.: Acta Chem. Scand. *9*, 1362 (1955); *10*, 887 (1956)
57. Reedijk, J., van Leeuwen, P. W. N. M., Groeneveld, W. L.: Rec. trav. chim. Pays-Bas *87*, 129 (1968)
58. Griffith, J. S.: The Theory of Transition-metal Ions, Cambridge University Press 1961
59. Condon, E. U., Shortley, G. H.: Theory of Atomic Spectra, Cambridge University Press 1953
60. Jørgensen, C. K., Schwochau, K.: Z. Naturforsch. *20 a*, 65 (1965)
61. Schäffer, C. E., Jørgensen, C. K.: J. Inorg. Nucl. Chem. *8*, 143 (1958)
62. Jørgensen, C. K.: Acta Chem. Scand. *10*, 500 (1956); *11*, 151 (1957)
63. Schröder, K. A.: J. Chem. Phys. *37*, 2553 (1962)
64. Jørgensen, C. K., Preetz, W., Homborg, H.: Inorg. Chim. Acta *5*, 223 (1971)
65. Schläfer, H. L., Gausmann, H., Zander H. U.: Inorg. Chem. *6*, 1528 (1967)
66. Schäffer, C. E.: Proc. 8th Int. Conf. Coord. Chem., Vienna 1964, pp. 77–81
67. Brawer, S. A., White, W. B.: J. Chem. Phys. *67*, 2043 (1977)
68. Tischer, R. E.: ibid. *48*, 4291 (1968)
69. Andrews, L. J., Lempicki, A., McCollum, B. C.: ibid. *74*, 5526 (1981)
70. Adamson, A. W., Dunn, T. M.: J. Mol. Spectr. *18*, 83 (1965)
71. Walters, R. T., Adamson, A. W.: Acta Chem. Scand. *A 33*, 53 (1979)

72. Durville, F., Boulon, G., Reisfeld, R., Mack, H., Jørgensen, C. K.: Chem. Phys. Lett. *102*, 393 (1983)
73. Schäffer, C. E., Jørgensen, C. K.: Mat. fys. Medd. Dan. Vid. Selskab *34*, No. 13 (1965)
74. Schäffer, C. E., Jørgensen, C. K.: Mol. Phys. *9*, 401 (1965)
75. Schäffer, C. E.: Structure and Bonding *5*, 68 (1968); *14*, 69 (1973)
76. Schäffer, C. E.: Pure Appl. Chem. *24*, 361 (1970)
77. Glerup, J., Mønsted, O., Schäffer, C. E.: Inorg. Chem. *15*, 1399 (1976)
78. Brorson, M., Jensen, G. S., Schäffer, C. E.: J. Chem. Educ. *63*, 387 (1986)
79. Tanabe, Y., Kamimura, H.: J. Phys. Soc. Japan *13*, 394 (1958)
80. Forster, L. S., Rund, J. V., Fulcaro, A. F.: J. Phys. Chem. *88*, 5012; 5017 (1984)
81. Flint, C. D., Matthews, A. P.: J.C.S. Faraday Trans. *II 70*, 1307 (1974)
82. Forster, L. S., Mønsted, O.: J. Phys. Chem. *90*, 5131 (1986)
83. Pott, G. T., McNicol, B. D.: J. Luminescence *6*, 225 (1973)
84. Kane-Maguire, N. A. P., Conway, J., Langford, C. H.: J.C.S. Chem. Comm. 801 (1974)
85. Adamson, A. W.: J. Inorg. Nucl. Chem. *28*, 1955 (1966)
86. Gutierrez, A. R., Adamson, A. W.: J. Phys. Chem. *82*, 902 (1978)
87. Eriksen, J., Mønsted, O. Acta Chem. Scand. *A 37*, 579 (1983)
88. Stokowski, S. E., Schawlow, A. L.: Phys. Rev. Lett. *21*, 965 (1968)
89. Glerup, J., Hodgson, D. J., Pedersen, E.: Acta Chem. Scand. *A 37*, 161 (1983)
90. Riesen, H., Güdel, H. U.: Chem. Phys. Lett. *133*, 429 (1987)
91. Pratt, G. W., Bailey, P. T.: Phys. Rev. *131*, 1923 (1963)
92. Allen, J. W.: Phys. Rev. Lett. *27*, 1526 (1971)
93. Gorkom, G. G. P. van, Henning, J. C. M., Stapele, R. P. van: Phys. Rev. *B 8*, 955 (1973)
94. Fournier, J. T., Landry, R. J., Bartram, R. H.: J. Chem. Phys. *55*, 2522 (1971)
95. Schläfer, H. L., Gausmann, H., Witzke, H.: ibid. *46*, 1423 (1967)
96. Adamson, A. W.: J. Chem. Educ. *60*, 797 (1983)
97. Jørgensen, C. K.: J. Physique (Colloq.) *46-C7*, 409 (1985)
98. Orgel, L. E.: J. Chem. Phys. *23*, 1824 (1955)
99. Oelkrug, D.: Structure and Bonding *9*, 1 (1971)
100. Reinen, D., Friebel, C.: ibid. *37*, 1 (1979)
101. Warren, K. D.: ibid. *57*, 119 (1984)
102. Jørgensen, C. K.: Chimia *27*, 203 (1973); *28*, 6 (1974)
103. Jørgensen, C. K.: Structure and Bonding *22*, 49 (1975)
104. Hathaway, B. J.: ibid. *57*, 55 (1984)
105. Boulon, G., Jørgensen, C. K., Reisfeld, R.: Chem. Phys. Lett. *75*, 24 (1980)
106. Mingardi, M., Porter, G. B.: J. Chem. Phys. *44*, 4354 (1966)
107. Wölpl, A., Oelkrug, D.: Ber. Bunsenges. *79*, 394 (1975)
108. Viane, L., D'Olieslager, J., Ceulemans, A., Vanquickenborne, L. G.: J. Am. Chem. Soc. *101*, 1405 (1979)
109. Condrate, R. A., Forster, L. S.: J. Chem. Phys. *48*, 1514 (1968)
110. Kirk, A. D., Ludi, A., Schläfer, H. L.: Ber. Bunsenges. *73*, 669 (1969)
111. Kirk, A. D., Schläfer, H. L., Ludi, A.: Canad. J. Chem. *48*, 1065 (1970)
112. Ludi, A., Güdel, H. U.: Helv. Chim. Acta *51*, 1762 (1968)
113. Schläfer, H. L., Wagener, H., Wasgestian, F., Herzog, G., Ludi, A.: Ber. Bunsenges. *75*, 878 (1971)
114. Eisenstein, J. C.: J. Chem. Phys. *34*, 1628 (1961)
115. Orgel, L. E.: Nature *179*, 1348 (1957)
116. Brorson, M., Schäffer, C. E.: Inorg. Chem., in press
117. Jørgensen, C. K.: Quimica Nova (São Paulo)
118. Jørgensen, C. K.: Chimia *42*, 21 (1988)
119. Greenberg, E., Katz, G., Reisfeld, R., Spector, N., Marshall, R. C., Bendow, B., Brown, R. N.: J. Chem. Phys. *77*, 4797 (1982)
120. Knochenmuss, R., Reber, C., Rajasekharan, M. V., Güdel, H. U.: ibid. *85*, 4280 (1986)
121. Liehr, A. D.: J. Phys. Chem. *67*, 1314 (1963)
122. Hammerling, P., Budgor, A. B., Pinto, A. (eds.): Tunable Solid State Lasers, Springer Series in Optical Sciences *47* (1985)
123. Holzrichter, J. F.: Nature *316*, 309 (1985)
124. Holzrichter, J. F., Campbell, E. M., Lindl, J. D., Storm, E.: Science *229*, 1045 (1985)

125. Crookes, W.: Proc. Roy. Soc. (London) *40*, 280 (1886); *42*, 25 (1887)
126. Targos, W., Forster, L. S.: J. Chem. Phys. *44*, 4342 (1966)
127. Blasse, G.: Structure and Bonding *26*, 43 (1976)
128. Weber, M. J.: Phys. Rev. *157*, 262 (1967); *B8*, 54 (1973)
129. Chatterjee, K. K., Forster, L. S.: Spectrochim. Acta *20*, 1603 (1964)
130. Goldsmith, G. J., Shallcross, F. V., McClure, D. S.: J. Mol. Spectr. *16*, 296 (1965)
131. Van Die, A., Blasse, G., Van der Weg, W. F.: J. Phys. *C18*, 3379 (1985)
132. Ghazzawi, A. M., Tyminski, J. K., Powell, R. C., Walling, J. C.: Phys. Rev. *B30*, 7182 (1984)
133. Suchocki, A., Gilliland, G. D., Powell, R. C.: ibid. *B35*, 5830 (1987)
134. Neuhaus, A.: Z. Krist. *113*, 195 (1960)
135. Bruce, A., Capobianco, J. A., Cormier, G., Krashkevich, D., Simkin, D. J.: J. Physique (Colloq.)
136. Bergin, F. J., Donegan, J. F., Glynn, T. J., Imbusch, G. F.: J. Luminescence *36*, 231 (1987)
137. Glerup, J.: Acta Chem. Scand. *26*, 3775 (1972)
138. Heber, J., Platz, W.: J. Luminescence *18*, 170 (1979)
139. Ferguson, J., Oosterhaut, B. van: ibid. *18*, 165 (1979)
140. Reisfeld, R.: Inorg. Chim. Acta *140*, 345 (1987)
141. Reisfeld, R., Eyal, M., Buch, A., Ish-Shalom, M.: Chem. Phys. Lett., in press

Bonding and Structure of Water Molecules in Solid Hydrates.
Correlation of Spectroscopic and Structural Data

Heinz Dieter Lutz

Universität Siegen, Anorganische Chemie, Postfach 10 12 40, 5900 Siegen, F.R.G.

Solid hydrate research of the last fifteen years is critically evaluated with regard to bonding and structure of water molecules. This review focusses on new results of structure determination and infrared and Raman studies in terms of hydrogen bonding and other intermolecular bonding interactions, distortion and disorder of water molecules, intermolecular and intramolecular coupling and anharmonicity of water bands, isotopic effects, and phase transitions. The techniques used for structure determination and spectroscopic measurements of solid hydrates are discussed.

Structure and Bonding 69
© Springer-Verlag Berlin Heidelberg 1988

1 Introduction

This review article is concerned with the structure, bonding, and dynamic processes of water molecules in crystalline solid hydrates. The most important experimental techniques in this field are structural analyses by both X-ray and neutron diffraction as well as infrared and Raman spectroscopic measurements. However, nuclear magnetic resonance, inelastic and quasielastic neutron scattering, and certain less frequently used techniques, such as nuclear quadrupole resonance, electron paramagnetic resonance, and conductivity and permittivity measurements, are also relevant to solid hydrate research.

The earlier literature up until 1970 has been reviewed in an excellent article by Falk and Knop[1]. Meanwhile, the many new results have been calling for a more recent review. A great many additional hydrates were studied yielding more interesting insights into the bonding and structure of water molecules. No other molecule can form as many compounds as water. Solid hydrates therefore serve as model substances for studying structure and bonding features.

2 Experimental Techniques

2.1 X-Ray and Neutron Diffraction

Both X-ray and neutron diffraction methods are applied to determine the structure of crystalline solid hydrates. Because of the very small scattering cross-section of hydrogen atoms for X-rays it is much desirable to solve the crystal structure by means of neutron diffraction techniques.

Nevertheless, it is possible to determine the position of hydrogen atoms in solid hydrates by X-ray methods, e.g., from difference Fourier analyses, provided that the accompanying atoms are not too electron-rich and the crystals are of sufficient quality (as, for example, in the case of the structure determination of barium hydroxide and barium chloride hydrates[2, 3]). However, the OH bond lengths obtained are (systematically) too small by about 13 pm[4].

Neutron diffraction studies of solid hydrates are difficult due to the large diffuse, i.e., incoherent, scattering of hydrogen atoms, which strongly increases the background of the diffraction data. Therefore, deuterated specimens are frequently used for structure determination.

2.2 Nuclear Magnetic Resonance

Proton and deuteron nuclear magnetic resonance techniques are applied in solid hydrate research both for determining the hydrogen positions[5-7] and for studying the dynamic processes of the water molecules[7-13].

From broad-line NMR studies of single crystals, i.e., the angle dependence of the separation of the proton resonance doublet, magnitude and orientation of the p-p vectors of water molecules are obtained and from these the hydrogen positions can be computed[5-7]. The accuracy of this method, which was used to a larger extent in the period from 1960–1975, is quite good compared to neutron diffraction studies[6]. However, it fails in the presence of more than two different p-p vectors.

Dynamic processes such as fluctuational motions of water molecules can be studied from the spin-lattice relaxation times of proton nuclear resonance[8-10]. In the presence of highly mobile H atoms, the separation of the proton resonance disappears and the remaining signal sharpens as for isotropic liquids[11]. Furthermore, the frequencies of H_2O librations (see Sect. 4.4) can be computed[12, 13].

2.3 Infrared Spectroscopy

Conventional infrared spectra of powdery materials are very often used for studying solid hydrates in terms of sample characterization (fingerprints), phase transitions, and both structural and bonding features. For the latter objects mostly deuteration experiments are included. However, it must be born in mind that the band frequencies observed (except those of isotopically dilute samples (see Sect. 2.6)) are those of surface modes rather than due to bulk vibrations, i.e., the transverse optical phonon modes, and, hence, not favorably appropriate for molecular and lattice dynamic calculations.

Sample handling techniques involve the use of mulls of paraffin (nujol) and fluorolube, the latter for studying the OH and OD stretching mode region, as well as discs of alkaline halides, KBr, CsI, etc. Spectra of high quality can be easier obtained with discs than with mulls. However, there may be shifts of the peak maxima (in the most cases to higher wavenumbers) as large as 5 cm^{-1}, but usually less than 2 cm^{-1}. No such shifts have been reported for the spectra recorded from mulls. Sometimes, but less frequently than supposed formerly, chemical reactions with the alkaline halide of the discs or the window plates of the cell may occur, e.g., ion exchange reactions. Thus, in KCl discs of $CuCl_2 \cdot 2\,H_2O$ there is a very rapid formation of $K_2CuCl_4 \cdot 2\,H_2O$[14]. Additionally, solid solutions with the matrix material can arise, e.g., in the case of alkaline earth halate hydrates[15], so that spectra of matrix-isolated anions are obtained.

Single crystal studies of solid hydrates are scarce. There are two experimental procedures possible: (i) transmission spectra of thin crystal plates (see, for example, Refs. 16, 17) and (ii) reflection spectra of crystal faces[18, 19]. Using polarized infrared radiation, the species (symmetry) and other directional features of the water bands can be determined. In the case of reflection measurements, the true transverse and longitudinal optic phonon frequencies can be additionally computed by means of Kramers-Kronig analyses and oscillator fit methods, respectively. Both experimental techniques, however, are relatively difficult because of the lack of suitable monocrystals, the requirement of preparing sufficiently thin, i.e., <0.1 mm, crystal plates (except for studying overtone bands, see Sect. 4.2.6), and the efflorescence or absorption of water at the crystal surfaces. In favorable cases, thin sheets of orientated powdery material can be obtained[20, 21].

2.4 Raman Spectroscopy

Raman spectra of solid hydrates are recorded for similar applications as the infrared spectra, viz. sample characterization, phase transition studies, etc. Because of the need of only small crystals, i.e., as small as 0.2 mm in diameter[21a)], Raman single crystal studies are a technique commonly used for assigning water bands (see, for example, Refs. 21a–24).

Since novel gas lasers became available as powerful sources (and, very recently, highly sensitive diode arrays as multichannel detection equipment), Raman spectroscopy has become a very important method in solid hydrate research.

Other advantages of Raman spectroscopy are (i) that no sample handling is required – polycrystalline material (or single crystals) are given in small glass tubes –, (ii) that the band maxima are equivalent to the true phonon frequencies (in the case of i.r. allowed bands, both the transverse and the longitudinal optic modes are observed)[25)], and (iii) the high speed of recording with the novel multichannel detectors, which enable kinetic studies, e.g., for phase transitions, and Raman thermal analyses[26, 27)]. Drawbacks are the low scattering power of hydrogen atoms, the almost complete disappearance of the water bands frequently observed in the presence of strong H-bonds[28)], and the high costs of a complete Raman equipment.

2.5 Inelastic and Quasielastic Neutron Scattering

By means of inelastic neutron scattering techniques the vibrational modes of solid hydrates can be studied additionally to the infrared and Raman methods. The advantage of the neutron scattering technique is that there are no restricting selection rules as for the infrared and Raman spectra. In the latter cases, only zone center phonon modes, i.e., those with wavevector $\vec{q} \approx 0$, are observed. Thus, in the case of coherent inelastic neutron scattering, the whole dispersion curve of a phonon branch can be recorded. In solid hydrate research inelastic neutron scattering techniques are particularly applied in studying H_2O librations[29, 30)] (because of the large scattering cross-sections, i.e., intensities, of these vibrations) and H_2O translatory modes[31)].

Furthermore, neutron scattering techniques, viz. quasielastic neutron scattering, are a valuable tool for studying other dynamic processes in solid hydrate research, such as rotatory and translatory diffusion of water molecules and hydrogen atoms, respectively (see, for example, Refs. 10, 32, 33).

2.6 Isotope Dilution Technique

In the last fifteen years the isotope dilution technique has become a well-established method in solid hydrate research, especially for infrared and Raman studies. For this method, which was first described in the late sixties[1, 34–36)], hydrate or deuterohydrate samples containing small amounts of HDO molecules matrix isolated in the bulk H_2O or D_2O are investigated. The main advantage of studying so-called isotopically dilute samples is the fact that the vibrations of dilute HDO molecules do not exhibit any intermolecular coupling, which would cause various splittings and frequency shifts (see Sect. 4.1) and, hence, obstruct the spectra obtained. Furthermore, the intramolecular coupling of the stretching modes of HDO molecules can be neglected. Therefore, the OH and OD band frequencies of the dilute HDO molecules are a direct measure of the lattice potential in the various hydrogen positions of the structure (see Sects. 4.2.2–4.2.4).

3 Structure of Solid Hydrates

3.1 Types (Classes) of Solid Hydrates

Solid hydrate research covers various classes of chemical compounds, which possess different importance and practical use. There are stoichiometric hydrates[1] and those with varying water content as zeolitic hydrates. There are true hydrates and pseudohydrates[1]. The latter contain water as hydroxide or hydroxonium ions or as –OH and –H groups ("water of constitution"). The true hydrates with separable H_2O molecules ("water of crystallization") include inorganic salts, i.e., the so-called salt hydrates, hydrates of organic compounds, and the clathrates, as, e.g., the novel gas clathrates. This article is mainly concerned with the salt hydrates.

In the case of salt hydrates, the water of crystallization can be coordinated to metal ions, e.g., as ligands in aqua complexes, or to anions and other proton acceptor groups (by H-bonds), or it can be present as weakly bound solvate molecules ("lattice water"). Both isolated H_2O molecules are found in the structure and oligomeric or polymeric networks[1, 37, 38]. The latter are classified into three-dimensional frameworks (tectohydrates), two-dimensional layers (phyllohydrates), one-dimensional chains (inohydrates), and oligomeric groups (nesohydrates)[38].

3.2 Coordination and Bonding of Water Molecules

For the bonding of the water of crystallization primarily the first coordination sphere of the H_2O molecules must be considered, that is number and coordination of the proton acceptor groups and metal ions in the neighborhood of the water molecules. Various classifications are reported in the literature[1, 4, 38–40] in order to describe the environment of H_2O molecules in solid hydrates. The system mostly used is that of Chidambaram et al.[39], which classifies H-bonded water molecules with respect to their lone-pair coordination (see Table 1 and Fig. 1).

The different coordinations of the H_2O molecules observed in solid hydrates are due to the fact that water molecules can act both as proton donors and proton acceptors (via the two lone-pairs). The most frequent coordinations are tetrahedral surroundings, i.e., types A, B, E, G, and H (class 2) (see Table 1). Trigonal and trigonal-pyramidal coordinations such as types C, D, and F (class 1) and I, J, and K (class 1'), respectively, are less frequent. Coordination numbers higher than 4, e.g., trigonal bipyramidal coordination,

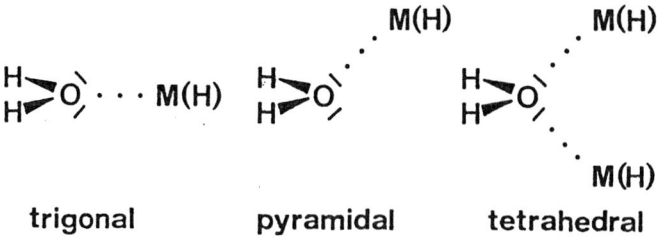

trigonal   pyramidal   tetrahedral

Fig. 1. Coordination of water molecules in solid hydrates (see Table 1)

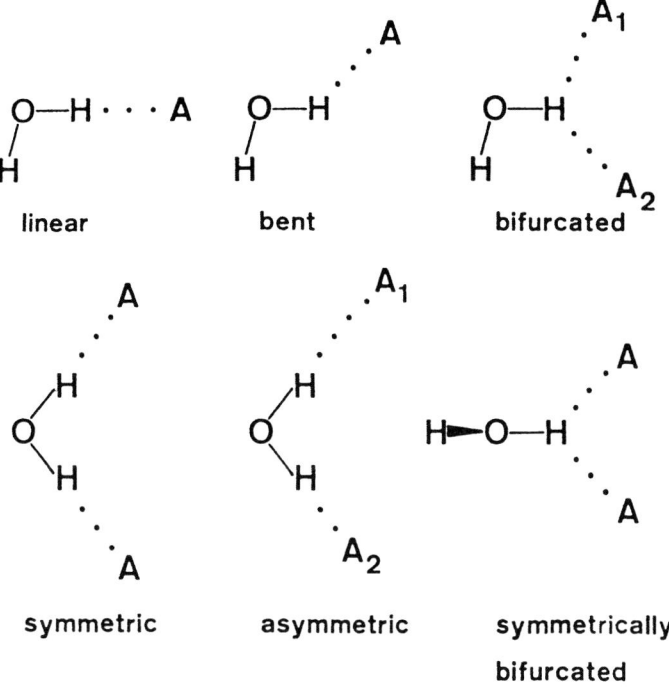

Fig. 2. Hydrogen bonds of water molecules in solid hydrates

are scarce[41, 42]. The mean volume occupied by a water molecule has been calculated as 0.0245 nm^3 [43].

Water molecules of crystallization are involved in more or less strong H-bonds. There are only few hydrates (or perhaps no: if bond lengths as large as the sum of van der Waals radii are allowed for very weak bonds) with non-hydrogen-bonded H_2O molecules, e.g., $Na_4SnS_4 \cdot 14\ H_2O$[44, 45] and $Na_2[Fe(CN)_5NO] \cdot 2\ H_2O$[44, 46]. The hydrogen bonds formed can be linear, bent, or bifurcated (see Fig. 2). However, H-bonds tend to be linear and

Table 1. Classification of water molecules in solid hydrates with respect to their lone-pair coordination (after Chidambaram et al.[39]) (see Fig. 1)

Type	Lone-pair coordination	Polyhedron	Class[41]
A	Lone-pairs toward two M^+	tetrahedral	2
B	Lone-pairs toward two M^{n+}	tetrahedral	2
E	Lone-pairs toward two H	tetrahedral	2
G	Lone-pairs toward M^+ and H	tetrahedral	2
H	Lone-pairs toward M^{n+} and H	tetrahedral	2
C	Bisectrix toward M^+	trigonal	1
D	Bisectrix toward M^{n+}	trigonal	1
F	Bisectrix toward H	trigonal	1
I	One lone-pair toward M^+	pyramidal	1'
J	One lone-pair toward M^{n+}	pyramidal	1'
K	One lone-pair toward H	pyramidal	1'

the acceptors tend to be located close to the water plane[42]. Especially with increasing bond strengths, deviations from linearity, i.e., OH⋯A angles $< 180°$, decrease[47, 48].

Furthermore, hydrogen bonds are called to be symmetric and asymmetric. However, these terms are used for different bonding features. (i) A hydrogen bond is said to be symmetric when the potential is of single-minimum type and there are equally strong bonds to the two atoms involved, i.e., O⋯H⋯O. (ii) A hydrogen bond is called symmetric if the donor and acceptor atoms are crystallographically equivalent, regardless of the type of potential curve[49]. (iii) A hydrogen bond (or a water molecule) is said to be symmetric when the two hydrogen atoms of the molecule are involved in equally strong H-bonds (see Sect. 4.2.2 and Fig. 2). In this review only the last definition is used.

The determined bond lengths vary with respect to both the acceptor groups (see Table 2) and the type of coordination of the water molecules[42]. Thus, for OH⋯O bonds, O⋯A (acceptor) distances are found in the range from 254.6 to 314.8 pm and H⋯A bond lengths from 152.0 to 225.8 pm, respectively[42]. The sum of van der Waals radii of H and O is assumed to be 240[42] and 260[1] pm, respectively. However, the chemical nature of the acceptor atoms, i.e., whether, for example, an O acceptor atom belongs to ClO_4^-, NO_3^-, SO_4^{2-}, PO_4^{3-}, OH^- etc., respectively[51-53], (see Sect. 4.2.3), and synergetic effects of bonds to metal ions are of great influence on the bond length and strength of hydrogen bonds[19, 53a-b]. Thus, increasing strength of bonding interactions with metal ions decreases the length of the H-bonds, e.g., in the case of $AlCl_3 \cdot 6\ H_2O$[19] and $Pb(XO_3)_2 \cdot H_2O$ (X = Cl, Br) (compared to the respective barium and strontium com-

Table 2. Average, minimum and maximum bond lengths (pm) and angles (°) observed for hydrogen bonds of water molecules[a] (after Chiari and Ferraris[42])

Acceptor		O	N	F	Cl
O–H[a]	Min.	89.1	90.2	93.0	93.0
	Av.	96.5	94.5	96.7	95.4
	Max.	102.9	96.8	100.1	99.0
O⋯A	Min.	254.6	281.7	256.3	308.6
	Av.	280.5	314.1	267.7	319.0
	Max.	314.8	341.9	290.6	332.1
H⋯A	Min.	152.0	186.7	165.1	212.8
	Av.	185.7	223.5	171.6	225.4
	Max.	225.8	254.0	194.4	238.1
Sum of van der Waals radii		240	250	235	280
H–O–H[a]	Min.	101.0	102.2	106.2	100.4
	Av.	107.2	103.9	108.1	106.3
	Max.	114.0	105.5	110.6	110.0
A⋯O⋯A	Min.	68.9	76.5	96.5	84.7
	Av.	108.3	95.9	103.6	102.8
	Max.	147.8	109.5	114.5	120.4
H⋯H	Min.	147.7	144.4	155.7	145.4
	Av.	155.5	148.6	156.9	152.4
	Max.	165.0	150.9	158.9	156.3

[a] in the gas phase, O–H 97.2 pm and H–O–H 104.5°[50].

pounds)[53a] (see Sect. 4.2.4). The shortest water H-bonds known have been observed for hydrates containing the very strong H-bond acceptor group OH$^-$ (see Sect. 4.2.3).

The lengths of the intramolecular O–H bonds of the water molecules, which, on the average, are shorter than in the gas phase, correlate with both the strength of the donated H-bonds and the type of coordination[42]. The bond lengths observed cover the range from 89.1 to 102.9 pm[42]. There are definite negative correlations between O–H and H$\cdots$A distances. The bond shortening in the case of weak hydrogen bonds is obviously caused by the repulsive part of the lattice potential. The H–O–H angles covering the range from 100.4 (in the case of $Sr(OH)_2 \cdot H_2O$ 99.9(6)°)[2, 54] to 114.0° are, on the average, about 2.5° larger than in the gas phase; this increase is correlated to the type of coordination (larger angles are found for trigonal surroundings), to the strength of the H-bond, and especially to the A$\cdots$O$\cdots$A angle between the acceptors[42].

Various models are reported in the literature[55, 56] for correcting the bond lengths and angles of H_2O molecules in solid hydrates for thermal motion and anharmonicity. However, as recently shown[56], both positive and negative terms exist, which partially compensate for each other. Therefore, data obtained by neutron diffraction do not show systematic errors larger than 3 pm and 2°, respectively, and, hence, the differences between the average water molecule in crystalline hydrates and that in the gas phase discussed above should be real[42].

The bonding interaction in solid hydrates gives rise to electron rearrangement in the H_2O molecules. Experimentally determined electron density maps unfortunately do not provide reliable information so far. Thus, especially in the lone-pair region of the O atom, the experimental density appears to be too low[57]. However, ab initio SCF calculations[57] for various model complexes have shown that different types of neighbors (H-bond donors or metal ions) give essentially the same electron redistribution in the H_2O molecules. There is an enhancement of the molecular dipole moment with electron depletion at the hydrogen sites and in the lone-pair region close to the oxygen nucleus and an extended region of slight electron excess between the oxygen and the neighboring metal ion or H-bond donor.

3.3 Orientational Disorder of Water Molecules

In the case of some hydrates, e.g., $SnCl_2 \cdot 2 H_2O$[58] and $MgSO_3 \cdot 3 H_2O$[21a, 59], the water molecules possess more than one possible orientations in the structure with different, energetically equal or unequal hydrogen positions ("orientational disorder" of the water molecules). The height of the energy barrier between these positions is in the range from 5 to 20 kJ mol^{-1}. According to this barrier and the thermal energy present, i.e., the temperature of the sample, there are frequent or less frequent rotational jumps of the water molecules between the two or more orientations possible. In the case of small energy barriers, so-called plastic crystals are formed with rotational diffusion of the water molecules, e.g., $LiI \cdot H_2O$[8] and $MgSO_3 \cdot 3 H_2O$[21a] (see also Sect. 4.2.5 and Ref. 60).

4 Correlation of Bonding and Structure with the Vibrational Modes (Band Frequencies, Intensities, and Halfwidths) of Water Molecules

4.1 Vibrational Modes of Water Molecules in Solid Hydrates – Intermolecular Coupling and Correlation Field Splitting of the Water Bands

Water molecules possess three internal vibrations, namely the symmetric (v_1, v_{sym}) and antisymmetric (v_3, v_{asym}) OH stretching modes and the H_2O bending vibration (v_2, δ). In solid hydrates there are, in addition, the external H_2O librations and H_2O translatory modes (lattice vibrations) (see Fig. 3). All these modes are more or less affected by the structural environment of the water molecules in the solid.

The main features that must be considered are the magnitude and symmetry of the static potential in the H_2O lattice site and intermolecular coupling (even between molecules of adjacent primitive unit cells) of the water bands. The former is shown by large frequency shifts of the water bands and decreased intramolecular coupling of the stretching vibrations (see Sect. 4.2.7) compared to those of free water molecules (see Table 3) and is discussed in Sects. 4.2–4.4 in more detail. The latter also produces frequency shifts and the so-called correlation field (Davydov) splitting of the water bands.

Because of the relatively large halfwidths of water bands, correlation field splitting has not been observed very often in the spectra of solid hydrates (except for frequency

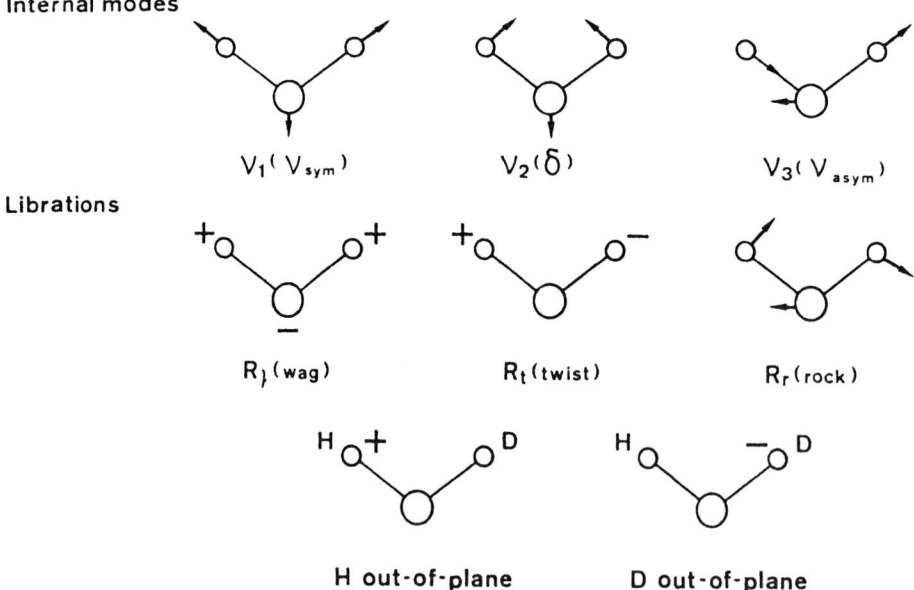

Fig. 3. Vibrational modes (internal modes and librations) of the water bands

Table 3. Energy range[a] of water bands (cm^{-1})

Solid hydrates	H_2O	HDO	D_2O
Stretching modes (v)	3600–3000	2600–2300[b]	2650–2300
Bending modes (δ)	1660–1590	1460–1400	1225–1175
Librations (R)	900– 350	900– 260[c]	680– 260
Translatory modes (T')	350– 100		330– 95
Free molecules[61, 62]	$v_1(v_{sym})$	$v_2(\delta)$	$v_3(v_{asym})$
harmonic (ω_e)			
H_2O	3832	1648	3943
HDO	2824	1441	3890
D_2O	2764	1206	2889
observed (ω_1)			
H_2O	3657	1595	3756
HDO	2727	1403	3707
D_2O	2671	1178	2788

[a] For hydrates with higher or lower band energies see Sects. 4.2.3, 4.2.4, 4.3, 4.4, and 4.5, respectively.
[b] Uncoupled $\bar{v}_{OD}$ of isotopically dilute HDO.
[c] Relatively large energy range because of the possibility of H and D out-of-plane motions (see Fig. 3).

shifts between the respective i.r. and Raman bands). Well-established examples are $BaBr_2 \cdot 2\ H_2O$[63] for stretching and bending modes and $SrCl_2 \cdot 2\ H_2O$[64] for librations. For a clear distinction whether band multiplets are due to correlation splitting or caused by crystallographically different H_2O molecules in the lattice, the isotope dilution technique (see Sect. 2.6) is recommended. In the former case the multiplets coalesce into a singlet on decoupling of the vibrations, in the latter one the multiplets remain upon deuteration.

4.2 OH Stretching Vibrations

In solid hydrates the stretching modes are shifted to lower wavenumbers compared to those of free H_2O molecules (see Table 3) for all compounds studied so far. Furthermore, the intensities and halfwidths of the respective i.r. and Raman bands alter characteristically (see Sects. 4.2.3 and 4.2.5). These findings, which are caused by the lattice potential, reveal valuable data on structure and bonding, especially on hydrogen bonds, and, hence, there are innumerable papers on this topic in the literature.

Such investigations are appropriately performed on spectra of isotopically dilute samples, especially those deuterated to about 5%, (see Sect. 2.6). Most studies carried out during the last decade used this technique. Infrared and Raman spectra of samples deuterated to about 5% are shown in Fig. 4. The OD and OH bands of HDO molecules present in samples deuterated to about 5 and 95%, respectively, are due to uncoupled, local modes (see Sect. 2.6); thus, the frequencies of the bands observed in the i.r. and Raman spectra must be equal (see Fig. 4).

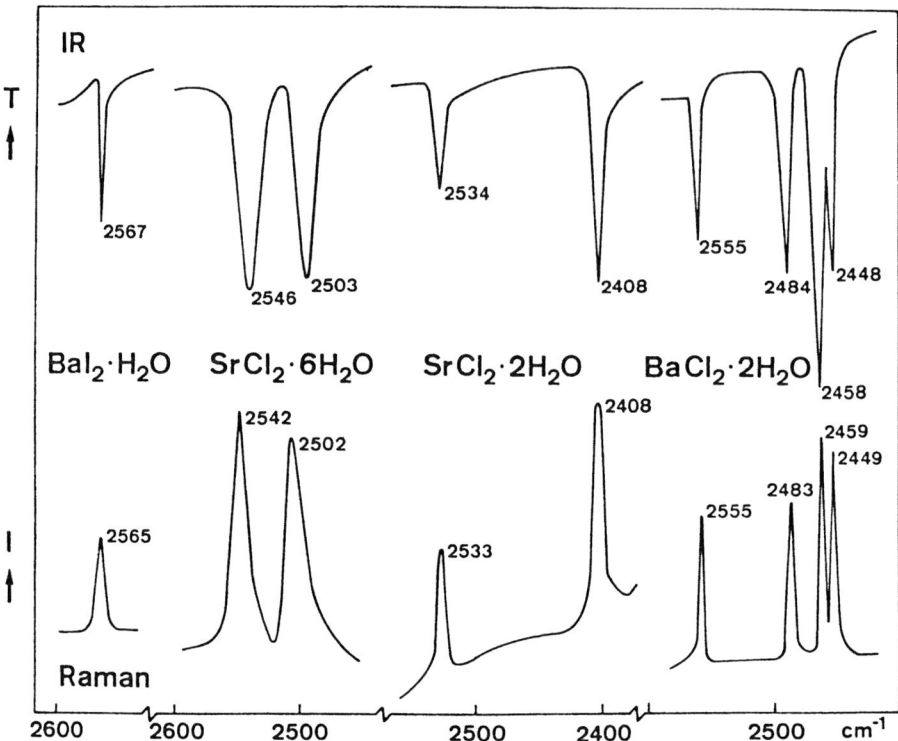

Fig. 4. Infrared and Raman spectra of isotopically dilute (5% deuteration) samples of $BaI_2 \cdot H_2O$ (one kind of symmetric H_2O)[65], $SrCl_2 \cdot 6\,H_2O$ (two kinds of symmetric H_2O)[66], $SrCl_2 \cdot 2\,H_2O$ (one kind of strongly distorted H_2O)[64, 66], and $BaCl_2 \cdot 2\,H_2O$ (two kinds of distorted H_2O)[67] at liquid nitrogen temperature (90 K)

The spectra of isotopically dilute samples mainly exhibit the following information, (i) number of different hydrogen positions (OH oscillators) in the structure, (ii) strength and arrangement of hydrogen bonds and other intermolecular interactions, (iii) distortion of the water molecules, and (iv) disorder of the hydrogen atoms. These topics are discussed first. Additionally, some features that can only be shown by means of the spectra of the neat compounds are reviewed.

The assignment of the two stretching modes is the same as for free water molecules, i.e., the symmetric mode (v_1) is found at lower wavenumbers than the antisymmetric one (v_3) as far as intramolecular coupling of the two OH vibrations occurs (see Sect. 4.2.2). However, there is evidence that the assignment of these modes must be reversed, i.e., $v_1 > v_3$, for compounds with very strong H-bonds, e.g., for $K_2C_2O_4 \cdot H_2O$[68] and $M(OH)_2 \cdot H_2O$ (M = Ba, Sr)[69] (see also Sect. 4.2.7 and Ref. 70).

4.2.1 Assignment of the OH (OD) Modes to the Various Hydrogen Positions (OH Oscillators) of the Structure

In the case where more than one or two crystallographically distinct H positions occur in the structure, e.g., with crystallographically different water molecules, the bands observed must be assigned to the various OH oscillators. In the past one has been using the correlation between the OH (and OD) stretching frequencies (ν_{OH}, ν_{OD}) of the H_2O and HDO molecules and the lengths of the hydrogen bonds present, e.g., r_{O-H}, $r_{H\cdots A}$, and $r_{O\cdots A}$, respectively, for this purpose (see, for instance, Refs. 1, 44, 71, 72). However, difficulties can arise for OH groups involved in equally or nearly equally strong H-bonds,

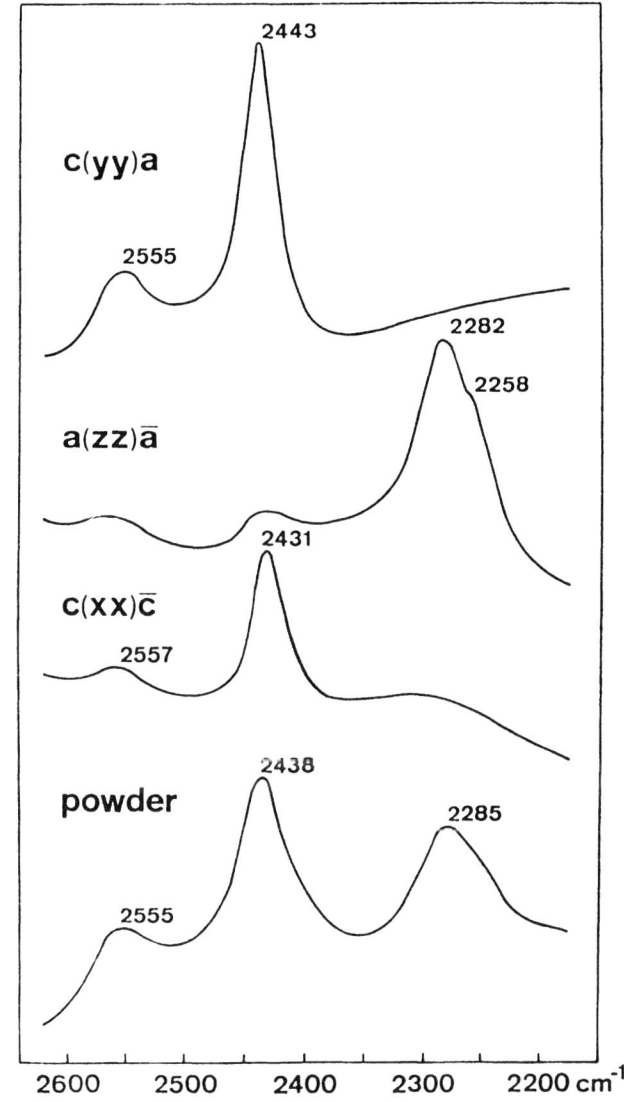

Fig. 5. Raman single crystal spectra of $MnSO_3 \cdot 3\,H_2O$ (5% deuteration) in the OD mode region[21a, 74a]. Five distinct OD oscillators (three crystallographically different H_2O molecules)[4] are present, those with bands at 2282 and 2258 cm^{-1} are positioned parallel to the c axis of the unit cell, that at 2431 cm^{-1} parallel to the a axis, that at 2443 cm^{-1} parallel to the b axis, and that at 2555 cm^{-1} in the ab plane of the structure

especially because the frequency vs. bond length correlations are inadequate in this case, as discussed in Sect. 4.2.3.

Very recently, single crystal studies of isotopically dilute samples were recommended for assigning the OH (and OD) bands of HDO molecules[59, 73–75]. These bands are strongly polarized and, hence, the intensities show directional behavior in the Raman spectra. The highest intensities are observed when the electric vectors of both incident and scattered light are oriented parallel to the O–H axis. For the Raman polarizability tensor of HDO molecules see also Refs. 76, 77. Thus, the spectra of $MgSO_3 \cdot 3 H_2O$[59], $MnSO_3 \cdot 3 H_2O$ (see Fig. 5)[21a, 74a], ice II[73], and $Sr(HCOO)_2 \cdot 2 H_2O$[74] were assigned by means of Raman single crystal studies. A similar technique, appropriate for assigning OH modes, probably is the infrared single crystal transmission study of isotopically dilute samples[78].

4.2.2 Distortion of the Water Molecules

The symmetry of free water molecules (C_{2v}) is normally not preserved in solid hydrates. Some examples with C_{2v}, C_2, or C_s^1 site symmetry of the water of crystallization are $CuCl_2 \cdot 2 H_2O$[79], $M(XO_3)_2 \cdot H_2O$ (M = Ba, Sr, Pb; X = Cl, Br, I)[15], and $M(OH)_2 \cdot H_2O$ (M = Ba, Sr)[2, 54]. In these cases, both H positions are equivalent and hence, only one OH and OD band is observed in the spectra of isotopically dilute samples (see Sect. 2.6). In the case of lower site symmetry, i.e., if the two H atoms are not equivalent (see Fig. 2), two OH and OD bands for each kind of water molecule are found in the spectra. The frequency difference of these two modes is a measure of the extent of the distortion of the H_2O molecules, which is caused by the different inter-molecular bonding (e.g., via H-bonds) of the two H atoms.

In the case of strong distortion, i.e., energy differences > 180 cm^{-1} for OH modes and > 130 cm^{-1} for OD modes, coupling of the two OH (OD) vibrations of the H_2O (D_2O) molecules to a symmetric and an antisymmetric stretching mode (v_1 and v_3) does not take place as for the OH and OD vibrations of HDO molecules. This was shown and established first by Schiffer and Hornig[80]. In this case, the frequencies of the two uncoupled bands resemble those of the H_2O (D_2O) bands in the spectra of the neat hydrates and deuterohydrates, respectively. The most distorted H_2O molecules known so far are found in $MgSO_3 \cdot 3 H_2O$ with $\Delta \bar{v}_{OD(OH)}$ 210(390) cm^{-1}[59]. In the case of undistorted, i.e., symmetrical, water molecules, the uncoupled HDO bands (v_{OH} and v_{OD}) (or the mean values of the two OH and OD bands of distorted molecules) are found intermediate between v_1 and v_3 of the H_2O and D_2O molecule, provided that there is no intermolecular coupling (see Sect. 4.1) of the water bands in the neat compounds.

The lattice site distortion of water molecules is further shown by the relative intensities of v_1 and v_3 of H_2O (D_2O) in the Raman spectra. In the case of not (or only little) distorted H_2O molecules, the Raman intensities of the antisymmetric stretching modes v_3 are very low[65].

4.2.3 Hydrogen Bonds

Water molecules are both fairly good proton donors and proton acceptors contrary to hydroxide ions[81], which are only poor proton donors, but very strong acceptors. Therefore, hydrogen bonds are present in liquid water and in the various polymorphic forms of ice as well as in all solid hydrates known, whereas in the case of solid hydroxides[28], compounds exist with hydrogen bonding, e.g., $Sr(OH)_2$, and also without hydrogen bonding, e.g., $Mg(OH)_2$ and $Ca(OH)_2$.

As known for a long time, in the presence of (and due to) hydrogen bonds, the OH stretching modes of the water molecules shift to lower wavenumbers – the reason is both the changed and additional force constants (see Sect. 4.6) and the increased anharmonicity of the vibrations (see Sect. 4.2.6) –, the halfwidths of the bands increase, and the intensities of the i.r. peaks grow. However, as shown only recently, the intensities of the respective Raman scattering peaks decrease with an increase in the strength of the H-bonds[28]. Thus, in the case of strong H-bonds, only very weak bands for all stretching and bending modes and librations are observed in the Raman spectra. For a further discussion of these findings see Ref. 21a.

The strength of the hydrogen bonds is correlated with the OH (and OD) stretching frequencies of the water molecules, particularly those of the HDO molecules in isotopically dilute samples (see Sect. 2.6). The strongest H-bonds, in which water molecules of solid hydrates are involved, are present in hydroxide hydrates because of the very strong proton acceptor strength of OH^- ions[81]. The bond length of such strong H-bonds is in the range of 255–280 pm for the $O \cdots O$ distances (see Sect. 3.2) and the OH (OD) stretching modes are observed at about $2800(2150)$ cm^{-1}. On the other hand, the weakest H-bonds found in solid hydrates are those of $NaClO_4 \cdot H_2O$ with OH (OD) modes (HDO of isotopically dilute samples) at $3584(2641)$ cm^{-1} [82].

The proton acceptor strength of the various entities present in solid hydrates can be arranged as follows $ClO_4^- < NO_3^- < ClO_3^- < BrO_3^- < IO_3^- < H_2O < SO_4^{2-} < SeO_4^{2-} < SO_3^{2-} < PO_4^{3-} < OH^-$ (see also Refs. 4, 51–53, 83, 84), $S_2O_3^{2-} < SbS_4^{3-} < AsS_4^{3-} < S^{2-}$ [85], and $PF_6^- < BF_4^- < I^- < Br^- < Cl^- < F^-$ [52, 65, 83]. The different acceptor strengths of the various proton acceptors is, at least partly, due to the different net charges of the acceptor atoms.

For the same acceptor, the strength of the H-bonds depends on the length and arrangement of the H-bonding. In the case of linear or nearly linear H-bonds (see Fig. 2) to H_2O molecules as acceptor groups in hydrates of metal sulfites[4], OD modes (HDO, uncoupled) range from 2503 cm^{-1} for $MnSO_3 \cdot 2 1/2 H_2O$ (weakest $HOH \cdots OH_2$ bond) to 2382 cm^{-1} for $CoSO_3 \cdot 3 H_2O$ o.-rh. (strongest bond). The respective $O \cdots O$ bond lengths are 299.8 and 272.8 pm. Thus, the strongest $HOH \cdots OH_2$ H-bonds in solid hydrates can be even stronger than those in ice (2421 cm^{-1}, 276 pm)[1, 71].

Experimental studies[4, 85a] as well as theoretical calculations[85b] indicate that hydrogen bonds are strengthened when water molecules act as both proton donor and proton acceptor ("cooperative effect"). Thus, an H_2O molecule bonded to another H_2O via oxygen lone-pair electrons behaves as a stronger acid than an isolated H_2O[85b].

Many ν_{OH} vs. bond length correlation curves are found in the literature[1, 44, 71, 72]. However, strong distinction has not been made in all cases for the various acceptor groups, the geometry of the H-bonds, frequency shifts due to intermolecular coupling (use of frequencies of coupled modes instead of uncoupled ones) etc. The best fits available have been given by Mikenda[44] (see Figs. 6 and 7).

Apart from linear H-bonds, bent and bifurcated hydrogen bond arrangements can be present in solid hydrates (see Fig. 2). As established by Falk et al.[86], the temperature dependence of the OH modes is different for linear and both highly bent or bifurcated

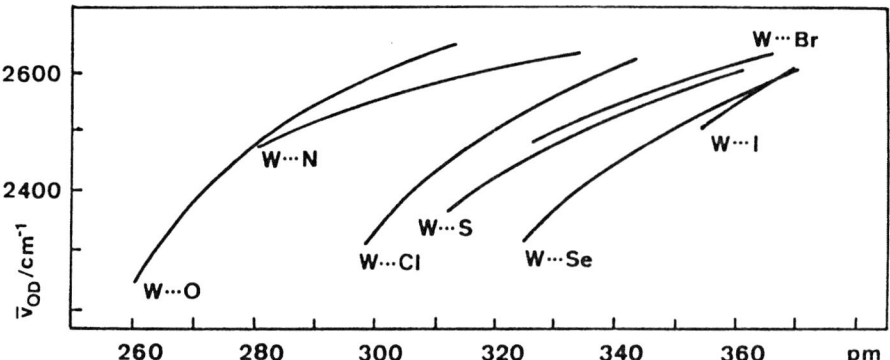

Fig. 6. $\bar{\nu}_{OD}$ energies (uncoupled HDO) vs $r_{O\cdots A}$ distances for various W $\cdots$ A hydrogen bonds in solid hydrates (after Mikenda[44])

H-bonds. In the former case, OH frequencies increase with increasing temperature, i.e., $d\nu_{OH}/dT > 0$, because the H-bonds are weakened as a result of the thermal expansion of the lattice. In the latter case, the temperature shift of the OH bands is negative, $d\nu_{OH}/dT < 0$, since such bonds are strengthened with increasing amplitude of the H_2O librational modes. However, for symmetrically bifurcated H-bonds, i.e., if all O, H, and the acceptor groups are located in one plane and the distances to the two acceptors are equal (see Fig. 2), a positive temperature coefficient (as in the case of linear bonds) must be produced[3]. This is, in fact, found, e.g., in $BaCl_2 \cdot H_2O$[65].

4.2.4 Other Intermolecular Bonding – Metal-Water-Interaction

The frequencies of the stretching modes of H_2O molecules are not solely influenced by the strength of hydrogen bonding. There are additional bonding features that must be taken into account, namely the interaction between the water molecule and adjacent metal ions[19, 44, 52, 65, 70, 85, 87–93a] and the repulsive part of the lattice potential[48, 53, 65].

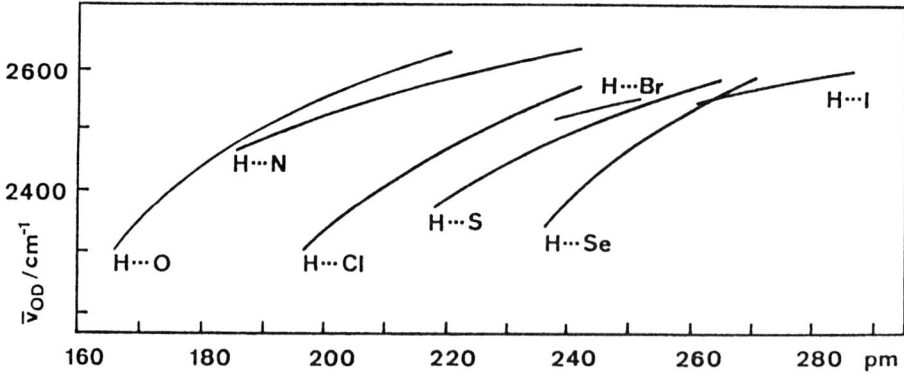

Fig. 7. $\bar{\nu}_{OD}$ energies (uncoupled HDO) vs $r_{H\cdots A}$ distances for various W $\cdots$ A hydrogen bonds in solid hydrates (after Mikenda[44])

Bonding interaction with metal ions both weakens the intramolecular OH bonds[89, 90, 92, 93] and increases the donor strength (acidity) of the hydrogen atoms and thus strengthens the H-bonds present ("synergetic effect")[19, 91, 93a] (see also Sect. 3.2). Both processes, especially the latter one, cause the OH bands to shift to lower wavenumbers than in the presence of H-bonds alone, and that nearly to an equally large amount[19, 44, 52, 70], i.e., up to 640 cm^{-1} [90]. The synergetic effect discussed above is shown from the relatively low-wavenumbered OH modes of hydrated salts of small, highly charged metal ions as Al^{3+} and those which tend to covalent bonding such as Pb^{2+} compared to salts of alkaline or alkaline earth metals with the same H-bond acceptor (see the discussions given in Refs. 19, 44, 93a) and from the spectra of isomorphous compounds with H-bond acceptors of different acceptor strength[65, 88].

Some hydrates are supposed to possess non-H-bonded water hydrogen atoms, e.g., $Na_4SnS_4 \cdot 14\ H_2O$ and $Na_2[Fe(CN)_5NO] \cdot 2\ H_2O$[44] with the highest – energy OD bands (HDO) at 2679[44] and 2653 cm^{-1} [93], respectively. In this case, the shift of the OH bands to lower wavenumbers – all solid hydrates known so far show OH frequencies smaller than those of free H_2O molecules (see Table 3) – must be caused by metal water interaction alone. However, the question arises whether weak hydrogen bonds are possible, even for H-bond lengths in the range of the van der Waals distances. Possibly then H-bonds cannot be excluded for any water of crystallization (see also Sects. 3.2 and 4.2.3).

In the case of solid hydroxides, however, it is confirmed that compounds with non-H-bonded OH$^-$ ions exist, e.g., most hydroxide hydrates[28, 94]. Thus, studies on the influence of bonding effects other than hydrogen bonding should be facilitated. Unfortunately, the vibrational energy of free OH$^-$, which was formerly assumed to be near 3700 cm^{-1} [94], was not exactly known. Both recent theoretical calculations[95] and estimates from the spectra of hydroxides with non-H-bonded hydroxide ions[94], however, resulted in values of about 3550 cm^{-1} for this mode, which is smaller than assumed earlier. Very recently, the vibration of free OH$^-$ was measured directly in the gas phase by means of photodetachment studies[96]. The value obtained was 3555.6 cm^{-1}. Because in solid hydroxides OH bands as large as 3700 cm^{-1}, e.g., for $Mg(OH)_2$[97], have been observed, metal-oxygen interactions obviously strengthen the intraionic OH bonds of hydroxide ions contrary to the result of metal oxygen interactions in the case of H_2O molecules (see the discussion given in Ref. 98).

The OH stretching modes of both water molecules and hydroxide ions (and most other polyatomic units) are additionally influenced, that is shifted to higher wavenumbers by the repulsive part of the lattice potential[48, 53, 93a, 94, 98]. This relatively small effect with maximum shifts of OH bands by about 30 cm^{-1}, can favorably be studied with isotopically dilute samples of hydrates containing matrix-isolated guest ions of smaller and larger size compared to those of the host ions. Such studies were recently performed with hydrates like $Ba(ClO_3)_2 \cdot H_2O$ containing BrO_3^- or IO_3^- guest ions[53]. The repulsive potential of the lattice and hence the induced frequency shifts become larger with increasing Coulomb (Madelung) energy per volume unit[15, 65].

4.2.5 Disorder of Water Molecules

In the case of static disorder of water molecules, for each position occupied by H atoms a distinct OH (and OD) band is observed in the spectra of isotopically dilute samples and, hence, static disorder can be established by vibrational spectroscopy if the two (or more) positions occupied by the disordered H atoms are crystallographically unequivalent. An example of static orientational disorder of H_2O molecules is $LiI \cdot H_2O$[99].

In the case of dynamical disorder, i.e., if the H_2O molecules reorient themselves rapidly (see Sect. 3.3), the halfwidths (defined as full width at half-maximum) of the OH bands become larger[58-60, 100-102]. For such studies the Raman peaks of the neat compounds or both the i.r. and Raman bands of uncoupled HDO molecules[21a] are suitable, but not the absorption peaks of conventional i.r. spectra because of band broading due to splitting of the transverse and longitudinal optic phonon modes[18, 19]. Examples reported in the literature are $SnCl_2 \cdot 2 H_2O$[58], $NiSiF_6 \cdot 6 H_2O$[100], $MgSO_3 \cdot 3 H_2O(H_2O \, III)$[21a, 59], $K_4[Fe(CN)_6] \cdot 3 H_2O$[101], and $Mg(ClO_4)_2 \cdot 6 H_2O$[102].

The temperature dependence of the halfwidths obeys an Arrhenius-type law, $\Delta v_{1/2} = A + B \exp(-V/kT)$[58, 60, 100, 101], i.e., $\Delta v_{1/2}$ grows strongly with increasing temperature. However, the activation energies thus obtained do not necessarily correspond to the barrier restricting the jumps between the different positions[58, 60]. This is due to the fact that band broadening and the temperature dependence can be caused also by other features, such as the violation of translation symmetry, anharmonic interaction with H_2O librations[103] or hydrogen bonding (see also Refs. 60, 100). In the case of Raman scattering, reorientational motion of H_2O is only involved in the halfwidth of the anisotropic part of the scattering tensor[100].

4.2.6 Mechanical Anharmonicity of OH Vibrations, Overtone Bands

Vibrations which include motions of hydrogen atoms are strongly anharmonic. Thus, the wavenumbers of OH stretching modes corrected for anharmonicity ω_e are about 200 cm^{-1} larger than those of the vibrational fundamentals ω_1 (see Table 3). In solid hydrates the anharmonicity of water bands strongly increases (with $\omega_e - \omega_1$ up to 600 cm^{-1})[106] as discussed below.

The magnitudes of the mechanical anharmonicities $x_e\omega_e$ are usually determined from the frequencies of the respective overtones ω_2, i.e., $x_e\omega_e = 1/2(2 \omega_1 - \omega_2)$. Because the

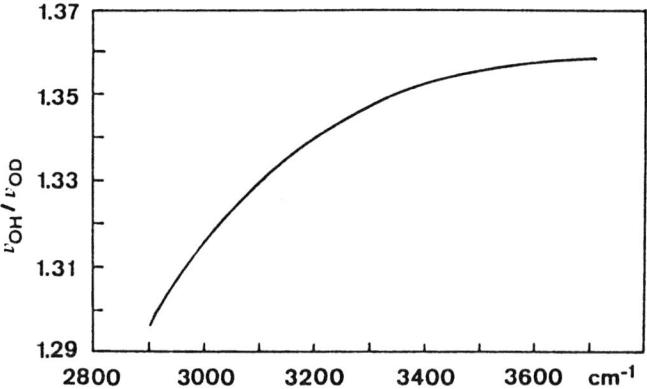

Fig. 8. Ratio $\bar{v}_{OH}/\bar{v}_{OD}$ vs $\bar{v}_{OH}$ of uncoupled HDO (after Berglund et al.[106])

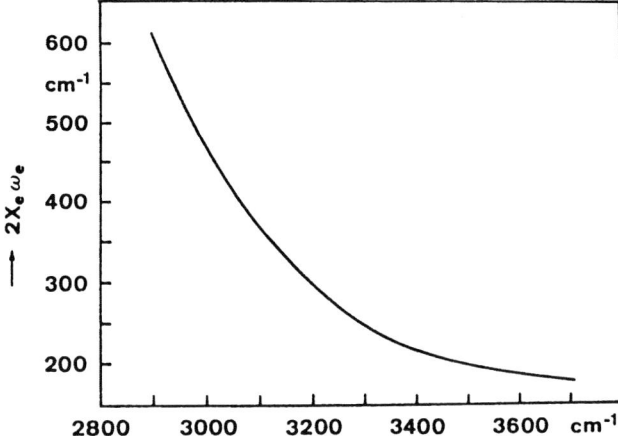

Fig. 9. Anharmonicity 2 $x_e\omega_e$ vs $\bar{v}_{OH}$ of uncoupled HDO (after Berglund et al.[106])

overtones of OH stretching modes are at about 7000 cm^{-1} (and thus beyond the range of conventional i.r. spectrometers) and the assignment of the bands observed in this spectral region is difficult, only little work has been done in this field, i.e., the near-infrared spectroscopy of solid hydrates (see, for instance, Refs. 75, 104, 105).

Approximate data of the anharmonicity of water bands, however, can be obtained from the isotopic shifts $\bar{v}_{OH}/\bar{v}_{OD}$ of the respective vibrational fundamentals ω_1, as shown by Berglund et al.[106]. Thus, the isotopic ratios $\bar{v}_{OH}/\bar{v}_{OD}$ strongly decrease with diminishing OH frequencies (see Fig. 8)[44, 106]. A further discussion of the reasons for the shift of the isotopic ratio is also given in Ref. 107. An alternative method of determining the anharmonicity of water bands from the respective fundamentals is described in Ref. 108.

The anharmonicity of the water bands strongly increases with increasing strength of the hydrogen bonds (see Fig. 9)[75, 106]. This behavior is entirely attributed to the higher amplitude of the stretching motion[108]. See also the discussion given in Ref. 109.

4.2.7 Intramolecular Coupling of the OH Vibrations

The intramolecular coupling of the two OH vibrations of a water molecule to a symmetric (v_1) and an antisymmetric (v_3) stretching mode is removed for strongly distorted H$_2$O molecules, where the separation between v_1 and v_3 has increased compared to that of free H$_2$O, viz. 99 cm^{-1}, (see Sect. 4.2.2 and Table 3). In many hydrates with symmetric water molecules – site symmetry C_{2v} (or an effective local potential of C_{2v} or nearly C_{2v} symmetry) – this separation is reduced (see also Sect. 4.2) to < 40 cm^{-1}, as in the case of CuCl$_2$ · 2 H$_2$O[79] and BaI$_2$ · H$_2$O[65].

The main reason for the reduced separation of the H$_2$O stretching modes has been ascribed to an increase of the stretching-stretching-interaction force constants F_{rr} (see Sect. 4.6) of the water of crystallization compared to the corresponding values for free H$_2$O molecules, viz. -0.10 (free H$_2$O) and -0.05 to $+0.12$ N cm^{-1} (solid hydrates)[70, 110, 111]. However, as recently shown[112], a reduced separation of v_1 and v_3 may also be due to the larger anharmonicity of the OH modes in the presence of strong hydrogen bonding (see Sect. 4.2.6). Finally the smaller intramolecular angle of the water molecule in some hydrates (see Sect. 3.2) is possibly correlated with both, the reduced separation and the reversed order of v_1 and v_3 (see Sect. 4.2) found for some of these compounds[69].

4.3 H_2O Bending Vibrations, Fermi Resonance

The bending mode $\delta_{H_2O}(v_2)$ of the water molecules in solid hydrates is mostly observed in the relatively small spectral range from 1590 to 1660 cm^{-1}. The highest and lowest wavenumbered modes found are at 1721 and 1582 cm^{-1}, respectively[113]. Because of this small spectral range, few structural data can be derived, apart from the fact that the appearance of bending modes is a proof of the presence of true hydrates (see Sect. 3.1). As claimed in the early literature and established in a recent study[113], the energies of the H_2O bending modes increase with increasing strength of the H-bonds, e.g., for $Sr(OH)_2 \cdot H_2O$ $\delta_{H_2O} = 1709$ cm^{-1} [28] (HOH angle 99.9(6)° [2, 54]), and decrease due to metal oxygen interaction, e.g., for $CuCl_2 \cdot 2\ H_2O$ $\delta_{H_2O} = 1587$ cm^{-1} [79]. Furthermore, the intramolecular angle of the water molecules (see Sect. 3.2) seems to be relevant in such a manner that the band energy becomes larger, if the angle decreases[21a, 69]. Site symmetry and extent of distortion of H_2O molecules obviously do not affect the bending modes to a larger extent.

The intensities of the H_2O bending modes, v_2, are usually smaller than those of the stretching bands, v_1 and v_3, in both the i.r. and Raman spectra of solid hydrates. However, there are hydrates with relatively low ($BaBr_2 \cdot 2\ H_2O$) and with high Raman intensity ratios ($BaCl_2 \cdot 1\ H_2O$)[114]. The intensities of the first overtones ω_2 of H_2O bending modes near 3200 cm^{-1} are often strongly increased due to Fermi resonance with OH stretching fundamentals. This possibility must be taken into account if bands in the OH stretching mode region have to be assigned. In the case of Fermi resonance between broad and intense H_2O stretching modes and relatively narrow overtone bands, negative absorption features, so-called Evans holes are observed[115], which sharpen at low temperatures.

In the case of strongly distorted water molecules (see Sect. 4.2.2), two HDO bending modes are observed in deuterated samples because of the two possible arrangements of the HDO molecule[86, 116]. Examples reported are $K_2SnCl_4 \cdot H_2O$[86], $Ba(NO_2)_2 \cdot H_2O$[116], and $Cs_2[Fe(CN)_5NO] \cdot H_2O$[17].

4.4 H_2O Librations

The librations (rotatory modes) of water molecules in solid hydrates (R) – hindered rotations around the three axes of inertia (see Fig. 3) – are normally observed in the relatively broad spectral region from 350 to 900 cm^{-1} [117]. The upper and lower limits found are 1123 ($CsOH \cdot H_2O$)[118] and 335 cm^{-1} ($BaI_2 \cdot H_2O$)[119]. The H_2O rotatory modes can be distinguished from other bands appearing in the spectral range under discussion, for example, internal modes of polyatomic ions, by deuteration experiments.

Because H_2O librations are very sensitive to the structural environment of water molecules, it was expected that these bands may reveal as much structural and bonding information as the stretching modes. However, all hitherto attempts (see Ref. 117) to analyse and interpret the H_2O rotatory modes of solid hydrates were not particularly successful and it seems that this state of the art will continue, at least in the near future.

The reasons for this failure are (i) the not fully solved problem of assigning the bands observed to the three possible rotatory motions, viz. the wagging (γ), the twisting (t), and the rocking (r) vibration (see Fig. 3)[117, 119–122], (ii) the not exactly known vibrational modes, i.e., the axes of rotation, especially in the case of distorted H_2O molecules (see Sect. 4.2.2)[1, 14, 119, 121, 123], (iii) the various intermolecular couplings of H_2O librations, in particular with translatory modes (lattice vibrations)[119, 121, 124] and internal vibrations of other polyatomic entities present[121, 123, 125, 126], and the extent of intramolecular coupling with the H_2O bending mode[121], and (iv) the simultaneous action of different external forces on the two hydrogen atoms and the oxygen atom. In a recent review[117], these issues are thoroughly discussed.

Several procedures are suggested in the literature (see Ref. 117) for the assignment of H_2O librations: (i) the different intensities of these vibrations in both i.r.[127] and Raman spectra[119, 123], (ii) the H_2O/D_2O and H_2O/HDO isotopic shifts[128, 129], (iii) trends obtained by force constant calculations[121, 130], (iv) Raman[21a] and i.r. single crystal studies. In special cases, coupling with internal vibrations of other present entities[123], and combination bands based on librations[105] give additional indications. For (i), the intensities of the respective i.r. bands should be $I_\gamma > I_r \gg I_t$, those of the Raman peaks $I_t > I_r > I_\gamma$ ((in the case of a tetrahedral environment of H_2O (see Fig. 1) $I_t \gg I_r > I_\gamma$)), because of the different changes in dipole moment and polarizability during the rotatory motions of the H_2O molecules. For (ii), the isotopic shifts, especially the H_2O/HDO shifts, should differ due to the different moments of inertia in such a manner, that ω_H/ω_D follows the order $R_\gamma < R_r < R_t$ (see Table 4). For (iii), after model calculations[121], the relative order of the H_2O librations is $\omega_\gamma > \omega_t > \omega_r$ for tetrahedrally coordinated H_2O molecules (types A, B, E, G, H)[39] (see Table 1) and $\omega_r > \omega_t > \omega_\gamma$ for water molecules in a trigonal arrangement (types C, D, F). These trends were also claimed in the early literature (see Refs. 1, 117, 129), but have not been understood so far. Apart from Raman and i.r. single crystal studies[21a], perfect assignments of the H_2O librations, however, are as yet not possible, especially in the case of both distorted water molecules and coupling of the librations with other vibrations.

The high sensitivity of H_2O librations to the structural environment and intermolecular bonding was analysed by several authors in spite of the numerous complications involved, as mentioned above. Thus, it has been claimed, but only proved for some limited series of hydrates, that the band frequencies expand with increasing metal-water interaction, as indicated by increasing $M–O_w$ stretching frequencies and decreasing $M–O_w$ distances[117, 119, 131], by the increase in the strength of H-bonds[28, 117, 119], and by the decrease in OH stretching frequencies[117, 119]. Because in twisting librations only the

Table 4. Isotopic shifts of H_2O librations[117, 121, 128, 129]

	$\omega_{H_2O}/\omega_{D_2O}$	$\omega_{H_2O}/\omega_{HDO}$	$\omega_{HDO}/\omega_{D_2O}$	$\omega_{H_2O}/\omega_{H_{oop}}{}^a$	$\omega_{H_2O}/\omega_{D_{oop}}{}^a$
R_γ (wagging)	1.339	1.089	1.230	0.90–1.05	1.25–1.45
R_t (twisting)	1.415	1.263	1.120		
R_r (rocking)	1.389	1.205	1.153	–	–

a $\omega_{H_{oop}}$ and $\omega_{D_{oop}}$, H and D out-of-plane motions (see Fig. 3).

hydrogen atoms, but not the oxygen atom, move, these modes are more appropriate for studying H-bonds than wagging and rocking librations[114, 119]. In the earlier literature some equations were derived in order to estimate the potential barrier heights of the water molecules from the torsional frequencies (twisting librations) (see, for example, Refs. 117, 122, 132–134). However, in this manner only approximate values can be obtained.

Some additional features of H_2O librations have also been discussed: (i) The bands under discussion possess relatively large halfwidths compared to other water bands, in the case of some hydrates even at liquid helium temperature as in $CuCl_2 \cdot 2 H_2O$[14]. This band broadening has been ascribed to dynamic disorder of the water molecules[14, 123, 134–135a]. However, violation of translation symmetry in the case of static disorder, e.g., for $CuCl_2 \cdot 2 H_2O$ at low temperature[14], must also be considered. From the temperature dependence of the halfwidths, activation energies for these dynamic processes have been calculated[136]. (ii) The temperature shifts of the band frequencies are normally negative, i.e., $d\omega_R/dT < 0$, due to the thermal expansion of the lattice. The positive temperature shifts of the librations found in some hydrates, e.g., for $SrCl_2 \cdot H_2O$[119], are discussed in terms of proton tunnelling[60].

The librations of HDO molecules (present in partially deuterated samples) exhibit splitting into two bands in the case of strongly distorted water molecules (due to the two possible arrangements HOD and DOH)[1, 129] in a similar way as do the stretching and bending modes (see Sects. 4.2.2 and 4.3). These splittings, first observed for $BaCl_2 \cdot 2 H_2O$[67], $SrCl_2 \cdot 2 H_2O$[64], and $Ba(NO_2)_2 \cdot H_2O$[116], are considerably larger, e.g., 98 cm^{-1} for $SrCl_2 \cdot 2 H_2O$, than calculated from the moments of inertia of distorted free HDO molecules, i.e, 20 cm^{-1} at the most. These findings confirm the results of force constant calculations[116, 121] that the librations of distorted HDO molecules – with two O(H, D) arms involved in differently strong H-bonds – are H and D out-of-plane motions rather than wagging and twisting librations. Normal twisting and wagging librations of HDO molecules are obviously only possible in the case of non-distorted water molecules (C_{2v}) in a tetrahedral environment (see Fig. 1)[14, 119]. In trigonal surroundings even non-distorted HDO molecules likewise reveal H and D out-of-plane motions rather than normal HDO librations as found, for instance, for $K_2CuCl_4 \cdot 2 H_2O$[1, 14, 120].

4.5 H_2O Translatory Modes (Lattice Vibrations)

The translatory vibrations of water molecules are mostly mixed with both H_2O librations and translatory modes of other entities present in the structure. Lattice vibrations containing H_2O motions are normally found in the spectral range from 100 to 350 cm^{-1}. In the case of pure T'_{H_2O} bands, the isotopic shift ratio $\omega_{H_2O}/\omega_{D_2O}$ is 1.054. However, pure T'_{H_2O} modes are rarely observed. Therefore, only little work was done on structural and bonding implications of H_2O translatory modes.

In the case of strong metal-oxygen interactions, i.e., bonding to highly charged metal ions such as Al^{3+} or some degree of covalency of the $M-O_w$ bond, the T'_{H_2O} bands are found at higher wavenumbers, e.g., at 536 and 485 cm^{-1} for $AlCl_3 \cdot 6 H_2O$[19] and $CuCl_2 \cdot 2 H_2O$[14], respectively. In these cases, the H_2O translatory modes are normally referred to as $M-O_w$ stretching bands.

4.6 Force Constant Calculations

In the past, force constant and normal coordinate calculations of water molecules were reported for all internal modes[20, 35, 68, 79, 110, 111, 130, 137], librations[116, 121, 122, 130, 138], and lattice vibrations[68, 121, 130, 139–141]. Studied hydrates are, for example, $CuCl_2 \cdot 2\,H_2O$[79, 110, 111], $MCl_2 \cdot 2\,H_2O$ (M = Mn, Fe, Co)[110], $SrX_2 \cdot 6\,H_2O$ (X = Cl, Br)[122, 141], $Ba(ClO_3)_2 \cdot H_2O$[68, 121, 142], $LiSO_4 \cdot H_2O$[20, 130], $CaSO_4 \cdot 2\,H_2O$[35, 111], $BeSO_4 \cdot 4\,D_2O$[140], and $Ba(NO_2)_2 \cdot H_2O$[116] (see also Table 5).

The various force fields used are based upon short-range forces in all cases studied. Therefore, the results of such calculations, especially those of lattice dynamical calculations, suffer from the fact that most solid hydrates are mainly ionic compounds with long-range electrostatic forces present as shown by the TO/LO splitting of the water bands (see Sects. 2.3, 2.4, and 4.2.5), which is found to be as large as 100 cm^{-1} for OH stretching vibrations[19]. Unfortunately, calculations have, to date, not been involving Coulomb forces as is the case, for instance, in the rigid ion or the polarizable ion model.

Nevertheless, normal coordinate model calculations of the H_2O librations, as reported recently[116, 121], are very valuable for understanding the vibrational modes, the mean square amplitudes (see also Ref. 144), and the relative order of these water bands (see Sect. 4.4). The force constants obtained, however, must be taken with caution (see the discussion given above).

The diagonal and off-diagonal force constants of the internal harmonic force field

$$2\,V = F_r(r_1^2 + r_2^2) + F_\alpha\alpha^2 + 2\,F_{rr}r_1r_2 + 2\,F_{r\alpha}(r_1 + r_2)\alpha\,,$$

and

$$2\,V = F_{r_1}r_1^2 + F_{r_2}r_2^2 + F_\alpha\alpha^2 + 2\,F_{r_1r_2}r_1r_2 + 2\,F_{r_1\alpha}r_1\alpha + 2\,F_{r_2\alpha}r_2\alpha$$

(for symmetric and distorted H_2O molecules, respectively) of H_2O molecules in solid hydrates vary significantly compared to those of free H_2O (see Table 5). There are various influences of both the external force field and the bond lengths and bond angle[137] of the H_2O molecule. The stretching force constants F_r decrease with increasing strength of the H-bonds[111]. (However, the low-frequency shift of the OH modes in the presence

Table 5. Intramolecular force constants of water molecules in the vapour and in some condensed systems (calculated from the harmonic frequencies ω_e) (see also Eriksson and Lindgren[111])

Compound	F_{r_1} [a]	F_{r_2} [a]	F_α [b]	F_{rr} [a]	$F_{r_1\alpha}$ [c]	$F_{r_2\alpha}$ [c]
$H_2O(g)$[110, 111]	8.454	–	0.697	–0.101	0.218	–
H_2O nitrogen matrix[111, 143]	8.338	–	0.698	–0.088	0.224	–
$NaPF_6 \cdot H_2O$[137]	7.991	–	0.703	–0.053	0.256	–
$Ba(ClO_3)_2 \cdot H_2O$[142]	7.724	–	0.684	–0.014	–0.0678	–
$Ba(ClO_3)_2 \cdot H_2O$[68]	7.80	–	0.68	–0.008	0.094	–
$CuCl_2 \cdot 2\,H_2O$[111]	7.39	–	0.70	0.060	0.23	–
$CaSO_4 \cdot 2\,H_2O$[111]	7.30	7.58	0.75	0.025	0.16	0.16
$K_2C_2O_4 \cdot H_2O$[111]	6.97	–	0.75	0.12	0.085	–

[a] N cm^{-1}; [b] $\times\,10^{-16}$ N cm rad^{-2}; [c] $\times\,10^{-8}$ N rad^{-1}.

of H-bonds is additionally caused by an increase in anharmonicity, especially in the case of strong H-bonds (see Fig. 9)[106, 111].) Furthermore, a linear relation was found between F_r and the stretching-stretching-interaction force constant F_{rr}, which increases as F_r decreases[110, 111] (see Sects. 4.2 and 4.2.7). From correlation field splitting of the water bands, intermolecular force constants can be derived[79, 145].

For interpreting the frequencies of the OH stretching modes, recently more sophisticated potential models including internal and external forces as well as harmonic, cubic, and higher terms were used[107, 146]. Furthermore, by means of *ab initio* calculations of clusters, such as $M–OH_2$ or $HOH \cdots X$, the frequency shifts of the water bands due to interaction with adjacent ions (see Sects. 4.2.3 and 4.2.4) could be verified[89, 90].

4.7 Isotopic Effects

Several deuterium isotopic effects are found in solid hydrates, additionally to the frequency shifts on deuteration due to the mass ratio (earlier discussed, Sects. 4.2.6, 4.3, 4.4, 4.5). These isotopic effects are caused by the different vibrational zero point energies, the different tunnelling probabilities, and the differently strong H-bonds of hydrates and deuterohydrates, i.e, stronger bonds for $OH \cdots A$ than for $OD \cdots A$ interactions[71].

The higher zero point energies and the larger strengths of H-bonds in the case of the non-deuterated compounds compared to the deuterated ones give rise to various structural, phase transition, and thermochemical isotopic effects. Thus, (i) the unit cell dimensions of deuterohydrates are somewhat larger than those of the non-deuterated specimens, the OD bond lengths are smaller, and the $OD \cdots A$ distances larger than the respective hydrogen distances (geometric isotope effects)[49], (ii) the temperatures of phase transitions are shifted[147], and (iii) the dehydration enthalpies of salt hydrates (which can be derived among others from the isotopic shifts of the stretching mode frequencies)[148] are smaller for the deuterated compounds than for the non-deuterated ones[148, 149]. In some cases, the deuterated compounds do not exist, e.g., in low temperature $M(SO_3)_2 \cdot 3 H_2O$ (M = Mg, Mn, Co)[21a, 59]. However, all these effects are small or even hardly detectable for most hydrates. Thus, geometric isotope effects are only observable in the case of strong H-bonds with $O \cdots O$ distances < 270 pm[49] or hydrates with dynamical disorder of the water molecules[21a]. On the other hand, in the case of solid hydroxides, such as CsOH, RbOH, and NaOH, the H/D isotopic effects discussed in general are relatively large and partly inverse to those of hydrates, e.g., stronger hydrogen bonds for $OD^- \cdots O$ than for $OH^- \cdots O$ interaction[150, 151].

Further H/D isotopic effects are (i) the increased intensities and decreased halfwidths of D_2O (and HDO) bands compared to those of H_2O in both the Raman and infrared spectra, and (ii) possible deviations from random distribution of H and D in partially deuterated specimens. From the relative intensities of the two OD (and OH) bands of HDO molecules in hydrates with strongly distorted water molecules (see Sect. 4.2.2) it is assumed that the hydrogen and deuterium atoms are not randomly distributed over the two H positions, but the deuterium atoms rather prefer those positions which are involved in stronger[66–66b] (weaker?)[74] H-bonds. For theoretical studies of the i.r. absorption and Raman scattering intensities of free H_2O, HDO, and D_2O see Refs. 77, 152.

4.8 Phase Transitions

The water bands of both i.r. and Raman spectra are most sensitive to even small structural changes. Therefore, vibrational spectroscopy is an even better tool for studying phase transitions of solid hydrates than most other experimental techniques, such as X-ray studies and thermal analyses (DTA, DSC). A great many studies of this topic have been reported (see, for example, Refs. 101, 102, 153).

In the case of phase transitions or other structural changes, the frequencies, the intensities, and the halfwidths of the water bands, especially those of the stretching and librational modes, as well as the temperature derivatives of these values change discontinuously. Hence, even very small phase transition features can be observed, such as disorder or small distortions of the coordinations spheres, often not detectable by X-ray methods, etc.

However, precaution must be taken to avoid over-interpretation of small changes in the band shapes. Thus, in the case of i.r. spectra, the band shapes observed are very sensitive to the type of preparation technique used for the samples, i.e., whether paraffin mulls or alkaline halide discs. The HDO bands of isotopically dilute samples (see Sect. 2.6) and the Raman spectra are less critical in this sense.

5 Concluding Remarks

Complementary investigations of solid hydrates by means of both spectroscopic methods, especially i.r. and Raman spectroscopy, and diffraction techniques enable detailed insight into structure, bonding, and dynamic processes of water molecules. There is no class of compounds which is more thoroughly studied in this respect than the crystalline hydrates.

The most important information obtainable pertains to (i) the number and lattice potential of the hydrogen positions present in the structure (Sects. 4.2.1 and 4.2.2), (ii) the extent of distortion of the water molecules (Sect. 4.2.2), (iii) the strength and arrangement of H-bonds and other intermolecular interactions (Sects. 3.2, 4.2.3, 4.2.4, 4.2.6), (iv) the distinction between true hydrates and pseudohydrates (Sects. 3.1, 4.3), (v) the possible orientational disorder of water molecules and dynamic processes, such as rotational diffusion, involved therein (Sects. 2.2, 2.5, 3.3, 4.2.5, 4.4), and (vi) the phase transitions and other structural changes upon heating or cooling (Sect. 4.8).

Acknowledgement. Many thanks are forwarded to Dr. J. Henning for helpful discussions.

6 References

1. Falk, M., Knop, O.: Water, A Comprehensive Treatise (F. Frank Ed.), Plenum Press, New York, Vol. 2, 1973, p. 106ff.
2. Buchmeier, W., Lutz, H. D.: Z. anorg. allg. Chem. *538*, 131 (1986)
3. Lutz, H. D., Buchmeier, W., Engelen, B.: Acta Cryst. *B43*, 71 (1987)

4. Engelen, B.: Habilitationsschrift, Univ. Siegen 1983
5. McGrath, J. W., Silvidi, A. A.: J. Chem. Phys. *34*, 322 (1961)
6. ElSaffar, Z. M.: ibid. *45*, 4643 (1966)
7. Sasidhar, Y. U., Rao, K. V. S. R.: J. Mol. Struct. *147*, 113 (1986); J. Phys. C.: Solid State Phys. *19*, 6241 (1986)
8. Chihara, H., Kawakami, T., Soda, G.: J. Magn. Reson. *1*, 75 (1969)
9. Seryshev, S. A., Vakhrameev, A. M., Afanas'ev, M. L., Kruglik, A. I., Bondarenko, V. S.: Yad. Magn. Rezon. Vnutr. Dvizheniya Krist. *1981*, 105 (1981)
10. Svare, I., Fimland, B. O., Otnes, K., Janik, J. A., Janik, J. M., Mikuli, E., Migdal-Mikuli, A.: Physica B + C *106*, 195 (1981)
11. Amm, D. T., Segel, S. L.: Z. Naturforsch. *41a*, 279 (1986)
12. McGrath, J. W.: J. Chem. Phys. *42*, 4111 (1965)
13. Afanas'ev, M. L., Kubarev, Yu. G., Zeer, E. P.: Yad. Magn. Rezon. Strukt. Krist. *1984*, 87 (1984)
14. Tanaka, H., Henning, J., Lutz, H. D., Kliche, G.: Spectrochim. Acta *43A*, 395 (1987)
15. Lutz, H. D., Christian, H., Eckers, W.: ibid. *41A*, 637 (1985)
16. Amalvy, J. I., Varetti, E. L., Aymonino, P. J., Castellano, E. E., Piro, O. E., Punte, G.: J. Crystallogr. Spectrosc. Res. *13*, 107 (1983)
17. Vergara, M. M., Varetti, E. L.: J. Phys. Chem. Solids *48*, 13 (1987)
18. Takahashi, H., Maehara, I., Kaneko, N.: Spectrochim. Acta *39A*, 449 (1983)
19. Wäschenbach, G., Lutz, H. D.: ibid. *42A*, 983 (1986)
20. Hayward, H. P., Schiffer, J.: J. Chem. Phys. *62*, 1473 (1975)
21. Wood, D. L.: ibid. *75*, 4809 (1981)
21a. Henning, J.: Thesis, Univ. Siegen 1988
22. Torres, A., Rull, F., DeSaja, J. A.: Spectrochim. Acta *36A*, 425 (1980)
23. Savatinova, I., Chisler, E. V., Rohleder, J., Jakubovski, B.: phys. stat. sol. (b) *101*, 233 (1980)
24. Choudhury, P., Ghosh, B., Raghuvanshi, G. S., Bist, H. D.: J. Raman Spectrosc. *14*, 99 (1983)
25. Lang, W., Claus, R.: J. Phys. Chem. Solids *44*, 789 (1983)
26. Lutz, H. D., Eckers, W., Christian, H., Engelen, B.: Thermochim. Acta *44*, 337 (1981)
27. Taylor, T. J., Dollimore, D., Gamlen, G. A., Barnes, A. J., Stuckey, M. A.: ibid. *101*, 291 (1986)
28. Lutz, H. D., Eckers, W., Schneider, G., Haeuseler, H.: Spectrochim. Acta *37A*, 561 (1981)
29. Thaper, C. L., Dasannacharya, B. A., Sequeira, A., Iyengar, P. K.: Solid State Commun. *8*, 497 (1970)
30. Kim, H.-J., Yoon, B.-G.: J. Korean Nucl. Soc. *11*, 1 (1979)
31. Boutin, H., Safford, G. J., Danner, H. R.: J. Chem. Phys. *40*, 2670 (1964)
32. Stahn, M., Lechner, R. E., Dachs, H., Jacobs, H. E.: J. Phys. C *16*, 5073 (1983)
33. Andersen, N. H., Kjems, J. K., Poulsen, F. W.: Phys. Scr. *25*, 780 (1982)
34. Hornig, D. F., White, H. F., Reding, F. P.: Spectrochim. Acta *12*, 338 (1958)
35. Kling, R., Schiffer, J.: Chem. Phys. Lett. *3*, 64 (1969)
36. Seidl, V., Knop, O., Falk, M.: Can. J. Chem. *47*, 1361 (1969)
37. Wells, A. F.: Structural Inorganic Chemistry, Clarendon Press, Oxford 1986, p. 659ff.
38. Bernal, J. D.: J. Chim. Phys. Physicochim. Biol. *50*, C1 (1953)
39. Chidambaram, R., Sequeira, A., Sikka, S. K.: J. Chem. Phys. *41*, 3616 (1964)
40. Hamilton, W. C., Ibers, J. A.: Hydrogen Bonding in Solids, Benjamin Inc., New York 1968, p. 204ff.
41. Ferraris, G., Franchini-Angela, M.: Acta Cryst. *B28*, 3572 (1972)
42. Chiari, G., Ferraris, G.: ibid. *B38*, 2331 (1982)
43. LeClaire, A., Monier, J. C.: ibid. *B38*, 724 (1982)
44. Mikenda, W.: J. Mol. Struct. *147*, 1 (1986)
45. Schiwy, W., Pohl, S., Krebs, B.: Z. anorg. allg. Chem. *402*, 77 (1973)
46. Bottomley, F., White, P. S.: Acta Cryst. *B35*, 2193 (1979)
47. Pedersen, B.: ibid. *B30*, 289 (1974)
48. Savage, H. F. J., Finney, J. L.: Nature (London) *322*, 717 (1986)
49. Ichikawa, M.: Acta Cryst. *B34*, 2074 (1978)
50. Cook, R. L., DeLucia, F. C., Helminger, P.: J. Mol. Spectrosc. *53*, 62 (1974)
51. Petrusevski, V., Soptrajanov, B.: J. Mol. Struct. *115*, 343 (1984)

52. Kristiansson, O., Eriksson, A., Lindgren, J.: VIIth International Workshop "Horizons in H-Bond Research", Marburg (FRG) 1985
53. Lutz, H. D., Henning, J.: J. Mol. Struct. *142*, 575 (1986)
53a. Lutz, H. D., Alici, E., Buchmeier, W.: Z. anorg. allg. Chem. *535*, 31 (1985)
53b. Ferraris, G., Fuess, H., Joswig, W.: Acta Cryst. *B 42*, 253 (1986)
54. Kuske, P., Engelen, B., Henning, J., Gregson, D., Lutz, H. D., Fuess, H.: Z. Kristallogr. (communicated)
55. Pedersen, B.: Acta Cryst. *B31*, 869 (1975)
56. Eriksson, A., Berglund, B., Tegenfeldt, J., Lindgreen, J.: J. Mol. Struct. *52*, 107 (1979)
57. Hermansson, K.: Acta Cryst. *B41*, 161 (1985)
58. Satija, S. K., Wang, C. H.: J. Chem. Phys. *68*, 4612 (1978)
59. Lutz, H. D., Henning, J., Buchmeier, W., Engelen, B.: J. Raman Spectrosc. *15*, 336 (1984)
60. Schaak, G.: J. Mol. Struct. *79*, 361 (1982)
61. Benedict, W. S., Yailar, N., Plyler, E. K.: J. Chem. Phys. *24*, 1139 (1956)
62. Nibler, J. W., Pimentel, G. C.: J. Mol. Spectrosc. *26*, 294 (1968)
63. Lutz, H. D., Pobitschka, W., Frischemeier, B., Becker, R. A.: J. Raman Spectrosc. *7*, 130 (1978)
64. Lutz, H. D., Pobitschka, W., Christian, H., Becker, R.-A.: ibid. *8*, 189 (1979)
65. Lutz, H. D., Christian, H.: J. Mol. Struct. *96*, 61 (1982)
66. Lutz, H. D.: Spectrochim. Acta *38A*, 921 (1982)
66a. Engdahl, A., Nelander, B.: J. Phys. Chem. *90*, 4982 (1986)
66b. Nelander, B.: XVIIIth European Congress on Molecular Spectroscopy, Amsterdam (Netherlands) 1987; J. Mol. Struct. (in press)
67. Lutz, H. D., Pobitschka, W., Frischemeier, B., Becker, R. A.: Appl. Spectrosc. *32*, 541 (1978)
68. Eriksson, A., Hussein, M. A., Berglund, B., Tegenfeldt, J., Lindgren, J.: J. Mol. Struct. *52*, 95 (1979)
69. Lutz, H. D., Kuske, P., Henning, J.: ibid. (in press)
70. Schiffer, J., Intenzo, M., Hayward, P., Calabrese, C.: J. Chem. Phys. *64*, 3014 (1976)
71. Novak, A.: Struct. Bonding *18*, 177 (1974)
72. Berglund, B., Lindgren, J., Tegenfeldt, J.: J. Mol. Struct. *43*, 179 (1978)
73. Minceva-Sukarova, B., Sherman, W. F., Wilkinson, G. R.: Spectrochim. Acta *41A*, 315 (1985)
74. Prieto, A. C., Rull, F., DeSaja, J. A.: J. Mol. Struct. *143*, 117 (1986)
74a. Lutz, H. D., Henning, J.: Z. Kristallogr. *178*, 148 (1987)
75. Savatinova, I., Anachkova, E.: Int. J. Quantum Chem. *29*, 1383 (1986)
76. Scherer, J. R., Snyder, R. G.: J. Chem. Phys. *67*, 4794 (1977)
77. John, I. G., Bacskay, G. B., Hush, N. S.: Chem. Phys. *51*, 49 (1980)
78. Keresztury, G.: J. Mol. Struct. *143*, 47 (1986)
79. Fifer, R. A., Schiffer, J.: J. Chem. Phys. *54*, 5097 (1971)
80. Schiffer, J., Hornig, D. F.: ibid. *49*, 4150 (1968)
81. Giguere, P. A.: Rev. Chim. Miner. *20*, 588 (1983)
82. Brink, G., Falk, M.: Can. J. Chem. *48*, 2096 (1970)
83. Kristiansson, O., Eriksson, A., Lindgren, J.: Acta Chem. Scand. *A38*, 613 (1984)
84. Buchmeier, W., Engelen, B., Lutz, H. D.: Z. Naturforsch. *41b*, 852 (1986)
85. Mikenda, W., Steidl, H.: Spectrochim. Acta *38A*, 1059 (1982)
85a. Kleeberg, H.: The Proceedings of the Interaction of Water in Ionic and Nonionic Hydrates (H. Kleeberg Ed.), Springer Verlag, Berlin, Heidelberg p. 89 (1987)
85b. Brakaspathy, R., Singh, S.: Chem. Phys. Lett. *131*, 394 (1986)
86. Falk, M., Huang, C.-H., Knop, O.: Can. J. Chem. *52*, 2380 and 2928 (1974)
87. Sartori, G., Furlani, F., Damiani, A.: J. Inorg. Nucl. Chem. *8*, 119 (1958)
87a. Zundel, G.: Angew. Chem. *81*, 507 (1969)
88. Lutz, H. D., Klüppel, H.-J., Feher, M., Bursian, S.: Ber. Bunsenges. Phys. Chem. *75*, 583 (1971)
89. Hermansson, K., Olovsson, I., Lunell, S.: Theor. Chim. Acta *64*, 265 (1984)
90. Falk, M., Flakus, H. T., Boyd, R. J.: Spectrochim. Acta *42A*, 175 (1986)
91. Kleeberg, H., Heinje, G., Luck, W. A. P.: J. Phys. Chem. *90*, 4427 (1986)
92. Varetti, E. L., Aymonino, P. J.: J. Mol. Struct. *79*, 281 (1982)

93. Amalvy, J. I., Aymonino, P. J.: Z. Phys. Chem. (Leipzig) 268, 15 (1987)
93a. Lutz, H. D., Henning, J.: The Proceedings of the Interaction of Water in Ionic and Nonionic Hydrates (H. Kleeberg Ed.), Springer Verlag, Berlin, Heidelberg p. 63 and 69 (1987)
94. Lutz, H. D., Eckers, W., Haeuseler, H.: J. Mol. Struct. 80, 221 (1982)
95. Werner, H. J., Rosmus, P., Reinsch, E. A.: J. Chem. Phys. 79, 905 (1983)
96. Owrutsky, J. C., Rosenbaum, N. H., Tack, L. M., Saykally, R. J.: ibid. 83, 5338 (1985)
97. Stanek, T., Pytasz, G.: Acta Phys. Pol. A52, 119 (1977)
98. Lutz, H. D., Henning, J., Haeuseler, H.: J. Mol. Struct. 156, 143 (1987)
99. Poulsen, F. W.: J. Raman Spectrosc. 17, 189 (1986)
100. Jenkins, T. E., Lewis, J.: Phys. Scr. 18, 351 (1978); J. Raman Spectrosc. 11, 1 (1981)
101. Savatinova, I., Anachkova, E., Ratajczak, H.: Raman Spectrosc., Proc. Int. Conf., 8th, 475 (1982); phys. stat. sol. (b) 91, 413 (1979)
102. White, M. A., Falk, M.: J. Chem. Phys. 83, 2467 (1985) and 84, 3484 (1986)
103. Brink, G.: Spectrochim. Acta 28A, 1151 (1972)
104. McCarthy, P. J., Walker, I. M.: ibid. 39A, 827 (1983)
105. Walker, J. M., McCarthy, P. J.: Can. J. Chem. 64, 1012 (1986)
106. Berglund, B., Lindgren, J., Tegenfeldt, J.: J. Mol. Struct. 43, 169 (1978)
107. Wojcik, M. J., Lindgren, J., Tegenfeldt, J.: Chem. Phys. Letters 99, 112 (1983)
108. Sceats, M. G., Rice, S. A.: J. Chem. Phys. 71, 973 (1979)
109. Sandorfy, C.: The Hydrogen Bond (P. Schuster, G. Zundel, S. Sandorfy Eds.), North Holland, Amsterdam, Vol. II, Chap. 13 (1976)
110. Fifer, R. A., Schiffer, J.: J. Chem. Phys. 52, 2664 (1970)
111. Eriksson, A., Lindgren, J.: J. Mol. Struct. 53, 97 (1979)
112. Mills, I. M., Robiette, A. G.: Molecular Physics 56, 743 (1985)
113. Falk, M.: Spectrochim. Acta 40A, 43 (1984)
114. Lutz, H. D., Christian, H., Haeuseler, H.: C.R.-Conf. Int. Spectrosc. Raman, 7th, Ottawa (W. F. Murphy Ed.), p. 120 (1980)
115. Othen, D. A., Knop, O., Falk, M.: Can. J. Chem. 53, 3837 (1975)
116. Eriksson, A., DeVillepin, J., Romain, F.: J. Mol. Struct. 140, 19 (1986)
117. Tayal, V. P., Srivastava, B. K., Khandelwal, D. P., Bist, H. D.: Appl. Spectrosc. Rev. 16, 43 (1980)
118. Lutz, H. D., Henning, J., Jacobs, H., Mach, B.: to be published
119. Lutz, H. D., Christian, H.: J. Mol. Struct. 101, 199 (1983)
120. Thomas, G. H., Falk, M., Knop, O.: Can. J. Chem. 52, 1029 (1974)
121. Eriksson, A., Lindgren, J.: J. Mol. Struct. 48, 417 (1978); Acta Chem. Scand. A32, 737 (1978)
122. Singh, B., Khanna, B. N.: Spectrochim. Acta 42A, 181 (1986)
123. Eckers, W., Lutz, H. D.: ibid. 41A, 1321 (1985)
124. Falk, M., Huang, C.-H., Knop, O.: Can. J. Chem. 53, 51 (1975)
125. Berenblut, B. J., Dawson, P., Wilkinson, G. R.: Spectrochim. Acta 29A, 29 (1973)
126. Soptrajanov, B., Ristova, M.: J. Mol. Struct. 115, 359 (1984)
127. Miyazawa, T.: Bull. Chem. Soc. Jpn. 34, 202 (1961)
128. Ichida, K., Kuroda, Y., Nakamura, D., Kubo, M.: Spectrochim. Acta 28A, 2433 (1972)
129. Lutz, H. D., Klüppel, H.-J., Pobitschka, W., Baasner, B.: Z. Naturforsch. 29B, 723 (1974)
130. Meshitsuka, S., Takahashi, H., Higasi, K.: Bull. Chem. Soc. Jpn. 44, 3255 (1971)
131. Tayal, V. P., Khandelwal, D. P., Bist, H. D.: Chem. Phys. Lett. 55, 136 (1978)
132. Sato, I.: J. Phys. Soc. Jpn. 20, 275 (1965)
133. Anachkova, E., Ratajczak, H., Savatinova, I.: phys. stat. sol. (b) 108, 65 (1981)
134. Canterford, R. P., Ninio, F.: Solid State Commun. 15, 1451 (1974)
135. Frindi, M., Peyrard, M., Remoissenet, M.: J. Phys. C13, 3493 (1980)
135a. Jain, Y. S., Singh, B., Khanna, B. N.: Pramana 18, 511 (1982)
136. Savatinova, I., Anachkova, E.: phys. stat. sol. (b) 82, 677 (1977) and (b) 84, 401 (1977)
137. Heyns, A. M.: Spectrochim. Acta 33A, 315 (1977)
138. Jain, Y. S.: Solid State Commun. 17, 605 (1975)
139. Corn, R. M., Strauss, H. L.: J. Chem. Phys. 76, 4834 (1982)
140. Pigenet, C.: J. Raman Spectrosc. 13, 66 (1982)
141. Donnelly, T. C., Nash, C. P.: Appl. Spectrosc. 36, 698 (1982)
142. Bertie, J. E., Heyns, A. M., Oehler, O.: Can. J. Chem. 51, 2275 (1973)
143. Fredin, L., Nelander, B., Ribbegard, G.: J. Chem. Phys. 66, 4065 (1977)

144. Eriksson, A., Hermansson, K.: Acta Cryst. *B39*, 703 (1983)
145. Hayward, H. P., Schiffer, J.: J. Chem. Phys. *64*, 3961 (1976)
146. Wojcik, M. J., Lindgren, J.: Chem. Phys. Lett. *99*, 116 (1983)
147. Tanaka, H., Negita, H.: Kagaku (Kyoto) *36*, 482 (1981)
148. Price, G. H., Stuart, W. I.: J. Chem. Soc. Faraday Trans. I *69*, 1498 (1973)
149. Tanaka, H., Negita, H.: Thermochim. Acta *41*, 305 (1980)
150. Bastow, T. J., Elcombe, M. M., Howard, C. J.: Solid State Commun. *57*, 339 (1986)
151. Jacobs, H., Mach, B., Lutz, H. D., Henning, J.: Z. anorg. allg. Chem. *544*, 28 (1987)
152. Swanton, D. J., Bacskay, G. B., Hush, N. S.: J. Chem. Phys. *84*, 5715 (1986)
153. Patel, M. B.: Solid State Commun. *53*, 431 (1983)

Author Index Volumes 1–69